Cotton

Resources Series

Peter Dauvergne & Jane Lister, *Timber*

Michael Nest, *Coltan*

Elizabeth R. DeSombre & J. Samuel Barkin, *Fish*

Jennifer Clapp, *Food* 2nd edition

David Lewis Feldman, *Water*

Gavin Fridell, *Coffee*

Gavin Bridge & Philippe Le Billon, *Oil*

Derek Hall, *Land*

Ben Richardson, *Sugar*

Ian Smillie, *Diamonds*

Cotton

Adam Sneyd

polity

First published in 2016 by Polity Press

Polity Press
65 Bridge Street
Cambridge CB2 1UR, UK

Polity Press
350 Main Street
Malden, MA 02148, USA

ISBN-13: 978-0-7456-8197-9
ISBN-13: 978-0-7456-8198-6(pb)

A catalogue record for this book is available from the British Library.

Library of Congress Cataloging-in-Publication Data

Names: Sneyd, Adam, 1978- author.
Title: Cotton / Adam Sneyd.
Description: Cambridge, UK ; Malden, MA : Polity Press, [2016] | Includes bibliographical references and index.
Identifiers: LCCN 2016002125| ISBN 9780745681979 (hardback : alk. paper) | ISBN 9780745681986 (pbk. : alk. paper)
Subjects: LCSH: Cotton trade--Political aspects. | Cotton--Economic aspects. | Cotton--Political aspects. | Economic development--Political aspects.
Classification: LCC HD9070.5 .S64 2016 | DDC 338.1/7351--dc23 LC record available at http://lccn.loc.gov/2016002125

Typeset in 10.5 on 13pt Scala by
Servis Filmsetting Ltd, Stockport, Cheshire
Printed and bound in the UK by CPI Group (UK) Ltd, Croydon, CR0 4YY

For further information on Polity, visit our website: politybooks.com

Contents

Acknowledgments

As I drafted this book, I consistently consulted the other Resources Series titles. I am consequently grateful to Jennifer Clapp, Peter Dauvergne, Jane Lister, Gavin Fridell, Derek Hall, Ben Richardson, Ian Smillie and to all of the other Resources authors.

That said, I bear sole responsibility if the content herein in any way fails to match the high standards that they have set. I am also highly indebted to all of the individuals that anonymously and keenly reviewed this project at the outset and prior to publication. At the University of Guelph, Danielle Mihok provided impeccable research assistance during the early phases, and while I focused on the write up, Steffi Hamann expertly took the reins on our new research program on commodity politics in Central Africa. Many colleagues at Guelph and beyond also enabled this book through supporting the development of my academic career. And Lauren Sneyd pushed me hard to make the manuscript come together on my sabbatical while we were visiting St Francis Xavier University. Thanks are due to Michael Cvetanovic for the wonderful time that Lauren and I were able to spend in Nova Scotia at Eagle's Nest, Malignant Cove, and also to Michael Sloopka and Kyle Harloff. I would also like to thank the members of the WTO Secretariat who helped me to engage directly with several high-level personalities involved in the global governance of cotton. Finally, at Polity, Pascal Porcheron and Nekane Tanaka Galdos were excellent shepherds, and Louise Knight's impeccable editorial expertise and patience allowed this book to see the light of day.

CHAPTER ONE

Spinning a Fibrous Tale

Resources like cotton are all about economics. Right? Think about it. Many daily newspapers and weekly news magazines give us good reason to believe that commodities are simply a humdrum piece of the economic system that enables modern consumer life. In print publications, stories about resources are often buried deep at the back of the business section. Online, the story tends to be much the same: typically, you have to go digging to find out what's going on in gold. Neoclassical economists, for their part, have convinced many media consumers and aspiring students that economics is the best discipline through which commodities like uranium can be understood. This radioactive viewpoint has had a significant half-life. And the fallout is evident every day on campuses around the world where or when students assert that abstract mathematical models are the sole route to "hard" knowledge about oil, pork bellies or other greasy commodities.

And the good news is that it is relatively easy to cut through this slick perspective. As commodity prices, resource exploration and investments boomed in the early 2000s, global resources became big academic business. Thereafter, when the global commodity bubble burst in 2014, an army of analysts stood at the ready to clean up the mess. Part of this clean-up crew rejected conventional neoclassical views on resource economics. They argued that analysts needed to foreground resource politics. In this light, commodity politics was seen to be too consequential to cover up or assume away. In putting politics at the center of their analyses, researchers working on raw materials as diverse as coffee and diamonds embraced a foundational premise of classical political economy. Their works – including all contributions to Polity's Resources Series – underscore the fundamental inseparability of politics and economics at all levels. Under this influence, political analysts in the clean-up crew do not consider the economics of global resources in isolation from their domestic, international, or global politics. Here, the shared understanding is that politics were as much a part of the story of guano in nineteenth-century Peru as they are of coltan in present-day Democratic Republic of Congo.

A big part of the challenge of wresting control over serious analysis and commentary on commodities from neoclassical economics relates to the language we still use to describe these goods. As a descriptor, perhaps "raw materials" best captures what people intend to do with the stuff that is used to make up other products. As such, on purely economic grounds, it makes sense that we continue to refer to bulk quantities of undifferentiated raw materials as primary commodities. But in drawing our focus only to those standard characteristics, the language we use every day can obscure more than it reveals. Through directing our attention solely to these goods themselves, users and abusers of the term "commodities" can effectively sanitize or "disappear" the processes that underpin the production of materials as diverse as timber and sugar. There are a great many companies and individuals around the globe that have deep interests in presenting commodity "stories" to be primarily about market movements. Up-to-date knowledge about financial and commercial developments related to the lifeblood of industrial business-as-usual is assuredly a corporate essential. But the raw deals that plantation workers and forests have reaped are equally a part of the commodity politics status quo. In the interest of getting those broader stories out more often, it might be helpful if we all took a cue from the slogan of Cotton Incorporated, the US-based cotton promotion agency. After all, resources or commodities are absolutely the "fabrics of our lives."

And the curious thing about cotton is that this fabric of our lives has flown largely under the radar. Unlike tea or peanuts, few people get to see the unadulterated raw stuff. There is also little doubt that most fans of the hit 1990s sitcom *Seinfeld* are blissfully unaware that Julia Louis-Dreyfus, one of the leads in that series, bears the family name of the leading global cotton merchant. Louis Dreyfus Commodities to this day remains a seemingly inoffensive family-run affair. And, truth be told, an extended discussion of the intricacies of that firm's daily operations could easily glaze the eyes of both ardent *Seinfeld* fans and critics alike the world over. But stay tuned: those activities are unquestionably coated in layers of political intrigue.

Sticking with another cultural product of the United States for a moment then – American college football – cotton has been similarly banal. The annual Cotton Bowl Classic had for years featured teams of marginal significance to the national rankings, and was only recently resuscitated by Goodyear, a tire and rubber company. In the heartland of the United States, this fiber has been so ostensibly bland that many Americans colloquially refer to confections spun from sugar as "cotton candy." In so doing, they have put a stamp of sweetness on a fiber with an otherwise unflattering flavor profile. Cotton candy does have a better ring to it than the reality – that this inedible fiber in raw form tastes atrociously bitter.

Activists, academics, and students have also contributed in their own small ways to the relative obscurity of cotton itself. Some of the biggest global campaigns of the past twenty years on issues linked directly to globalization and your T-shirts have missed this crop by a figurative "micronaire." The latter term is industry jargon for the measurement of just how thick the cell walls of cotton fibers are. Quality experts consider the thickness of the micronaire to be a key indicator of fiber quality. So, to be blunt, when it came to cotton, anti-sweatshop activism was then a little thin. Without a doubt, the hard work of labor activists to expose pay and workplace health and safety scandals in Indonesia, China, and elsewhere has been spot on and effective. In the 1990s, Jeffrey Ballinger and other leading lights in the clean clothes movement achieved fantastic results when they pushed the big branded clothing retailers to develop and introduce new standards for their suppliers.[1] And they did so once again in the aftermath of the factory fire and collapse scandals in Bangladesh, and also after many of the initial supplier codes failed to yield the hoped-for textile and garment worker benefits and protections.

But for every ounce of sweat shed by those devoted to making shop floors better for people and the planet, it can still seem that only a few drams have been shed in the service of making cotton work better for those that grow it. Given the persistent prominence of sweatshops in the global mediascape, that would be a reasonable surface-level impression. However, the reality revealed through research is strikingly different. High-level transnational political dramas have played out on cotton specifically over the past decades. This book will detail many of the consequential machinations and maneuvers that have recently changed the world order for cotton. Today, civil society groups, big businesses, and governments have latched onto the idea that the world needs cotton that is farmed and traded more ethically and sustainably. And a broadly shared understanding on how to best facilitate the realization of those noble objectives has not emerged. As such, there has been a profusion of high-stakes politics. This politics provides especially gripping material for political junkies because many industry insiders continue to deny its existence. Some directly reject the proposition that the cotton business is fundamentally political. And their denials are as political as they are ridiculous.

My task in writing this contribution to the Resources Series is to convince you that the geopolitics of cotton is anything but the stuff of a staid global commodity trade.[2] The transnational politics associated with this fiber is also not an impenetrable thicket. This tangled mass of high-level interrelationships and low-level chicanery can be unwound. And it should be in the interest of enhancing the uptake of fashionable and enduring principles, including informed global citizenship and ethical consumption.

Take for example some of the politics of cotton on the land. One of the biggest and most long-standing farmer dilemmas with this crop is enveloped in politics. The perennial question that hangs over every individual farmer and the global industry as a whole is rather straightforward. Should we plant cotton, or should we plant something else? Politics often bears directly on this choice where and when legislation or policy offers support to those that choose to grow it. It also comes into play in places where individual farmers do not get to make that choice directly. Owing to their status as tenants, workers or disempowered family members, sometimes the principal farmers of cotton tend not to be the ones that hold the power to make decisions about whether or not they should go with it.

The derivative questions linked to the choice to plant cotton are equally political. If we do plant cotton, how will we feed ourselves? And if we do not plant cotton, how will we feed ourselves? The idea of cottonseeds fried in cottonseed oil with a side of mashed cotton plant might have quirky appeal to comfortable people that do not confront this dilemma. But the food question is serious for those that face it. Farmers must be convinced that they can navigate the trade-off between cotton in the ground and their capacity to make food available or to access food at the market. For some producers with full bellies, this is a marginal consideration. But for the vast majority of those involved with cotton globally, there is no doubt that the food issue is overarching. And many powerful voices with an interest in keeping cotton on the land seek to influence farmer answers to the food question.

The politics of securing cotton for the country is perhaps better known. Many states that produce cotton, or that rely on it for industry, have for centuries attempted to control the fiber at home and abroad.[3] Where national interests have been linked to the availability of cotton, and business and high-level politics have become intertwined, many world-changing stocks and flows have emerged and become entrenched. For instance, when business interests in securing cheap cotton and furthering the industrial development of cotton were equated with the national interest in Britain, imports of finished fabrics from the Empire were curtailed. Subsequently, after the stock of textile firms was bolstered at home, British fabrics flooded the realm where the sun never set on British exports. Moreover, as demands for cheaper cotton from industry grew, the total global stock of enslaved people simultaneously rose. Cross-border flows of slaves and cotton fueled "economic growth" and industrialization in Britain, and then in the rest of Europe and the

United States. Along with sugar, cotton underpinned this hideous system. And controlling cotton for the country has also been a dirty business for "free" farmers at home. Many smallholders around the globe have suffered in the name of national industrial development. In a range of diverse places, cotton farmers have been seriously exploited. Over the course of the last century and down to today, many have been short-changed via the payment of artificially low prices.[4] Where governments have intervened to keep farm-gate prices low or businesses have colluded to reduce payouts, farmers have faced effective taxes on their outputs. And these "taxes" have sometimes lined the pockets of industry insiders. At other times, they have been used to subsidize the emergence of employment-generating spinning and weaving manufacturing operations of varying quality and durability.

On another front, cotton in company hands is a persistent source of political rhetoric and controversy. Corporations involved in the buying and processing of cotton to remove the seeds might seem to be pretty remote from big politics. At first glance, it could easily appear to be the case that company politics is of limited consequence beyond this industry. A political bribe here to get a standard shipping container there, for example. But the reality could not be more contrasting. Cotton company actions have considerable global political spillovers. For starters, domestic buyers and the transnational merchants that move cotton across borders are engaged in an activity that since the 1980s has become much more financially complex. Put simply, tools that enable companies to manage financial risks associated with cotton, such as futures contracts, have now become a really big part of this business. And the politics of futures and derivatives market regulation, or the lack thereof, and of financialization more generally, are matters of serious concern at the highest levels. Jennifer Clapp's must-read analysis of developments in this area appears in the Resources Series in her wonderful book *Food*.[5]

Beyond the politics of finance, company control over cotton has yielded negative externalities that have accelerated global environmental change.[6] The demand for ultra-white cotton that continues to emanate from some quarters has encouraged farming practices that, to put it mildly, have been far too intensive. As alternative approaches to cotton that rely less on agrochemicals have been successfully tested, companies that advance new ideas about cotton have emerged. These upstarts now challenge the old view that more chemicals and more water are necessary to grow quality fiber successfully. In the new order for cotton, companies can and do clash openly and politically over best practices.

Finally, the global cotton trade itself has animated renewed inter-state geopolitical conflict. Some states that export cotton and that are members of the World Trade Organization (WTO) have used WTO rules to challenge the cotton support policies of other WTO members.[7] Brazil and a group of African countries that depend on cotton have pushed back against US cotton policies that they have deemed unfair. The former launched a trade dispute and ultimately received compensation, and the latter grouping was able to inject the cotton trade issue straight into the heart of ongoing trade negotiations. Unfortunately for the African group, after over a decade of negotiations on this topic, culminating in December 2015, scant progress was made on their political demands for liberalization.

The inter-state dimensions of cotton trade politics are also bigger than the subsidy issue. As Europe's cotton policies have faced enhanced geopolitical scrutiny and Africa's exports of raw cotton lint have shifted away from Europe towards Asia, a more regionally oriented geopolitics has emerged. The European Union has financially supported the efforts African countries have made to work together to add more value to cotton in Africa. This so-called "road map" aims to help Africans do more together to spin cotton up, dye it, weave it, and piece it together as garments. As such, Africans are now using European support to challenge the dominance of Asian-based firms in the cotton value chain. Not to be outdone, China has scaled up its cotton production volumes. As well, its domestic textile investors have been encouraged to "go out" in search of lower-cost production locations. Moreover, in 2014 when China released some of its massive cotton stockpile, its foreign suppliers were largely cut off. This political move reduced the raw material costs of its overseas textile interests and of the global textile industry as a whole. And in the context of this renewed geopolitics, several high-profile scandals linked to the political exercise of control over the cotton trade have come to light.[8]

This politics has not elicited the kind of attention that gets paid to what clothing and bedding designers actually do with cotton, or what they do on their own time. To a certain extent this is understandable. Raw material politics tends only to hit the front pages when it is associated directly with death, destruction, or impending economic doom. For instance, fans of the classic James Bond movies of the 1970s would surely not think about cotton as being central to the politics of the plot lines. As Bond, Sir Roger Moore's numerous safari suits were a fashionable and enduring sidebar to all of that ticket-selling international intrigue. Today, the nightlife misadventures and errant tweets of leading fashionistas tend to get blanket social media coverage. Likes or shares of links related to their occasional commitments to source more "ethical" cotton, or simply to do better for the planet, have no hope of breaking the internet. But everybody still uses the stuff every day. If you did not dry off with a cotton towel this morning, you definitely will when you travel in search of your preferred paradise. If you can afford to do so. If not, cotton is omnipresent wherever you are, if you care enough to stop for a second to take a look.

Since this book is about the politics of cotton, I need to be straightforward with you about what you should not expect from it. First and foremost, this book is not intended to be a reference regarding cotton production and export volumes or trends. The secretariat of an inter-state, international organization – the International Cotton Advisory Committee (ICAC) – has been tasked with publishing detailed literature on those specifics on a regular basis. Similarly, if you are interested in reading extended analyses of price developments, market conditions, or trends, you will not find much in here. In that case, make sure that you check out the public access sections of the online Cotton Outlook market information subscriber service. Or get your credit card out to dig deeper into their database.

And if you really want to know more about the specifics of cotton policy in all producing and exporting countries, internet search engines should be your primary go-to. This applies doubly for those of you who might have or develop an interest in organic or fair-trade cotton standards, or in where you can purchase ethical or sustainable fashion online. Additionally, plenty of fantastic academic books and articles in scholarly journals on cotton have recently come out. So if you are itching to delve into the history of cotton in the world system, learn more about processes that aim to assure the quality of the fiber, or study the economics of the business in particular places, check out the Selected Readings that appear before the index. Just do not expect this book to engage too much with arcane academic debates.

You have in your hands a book on cotton that aims to push you to think about politics. To do so, it presents some theory out of necessity; hopefully in accessible language that has been sexed up enough to hold your attention. If it ever does not, please feel free to put it down and pursue your own exploration of what is going on under the cover of cotton in your neighborhood. When you are done, the analytical framework that I offer up will still be right here. This framework aims to impart the fundamentals of who gets to control cotton, where, and when. It directs attention to the principal ideas, institutions, and power relations that have shaped the status quo for cotton. Since the geopolitics of cotton is dynamic, this framework also enables us to grasp some of the forces and factors that facilitate change or create path dependencies in the global governance of this commodity.

Some might claim that this orientation is overly value laden. That a focus on politics can amount to no more than the politicized selection and presentation of biased information. But such an appraisal would neglect the gulf between polemics and the really "hard" science of political analysis. At the outset, let me assure you that industry self-promotion materials and the research that keeps cotton moving presuppose an enormous overarching value. They are weaved up with the fiber and fundamentally wedded to it. Consequently, it would be appropriate to consider this analytical book to offer a corrective or a "counter-narrative" on cotton.

As I see it, the governance of global resources is fundamentally about how states and non-state actors, working together or separately and at multiple levels, exercise control. It is about the exertion of direct influence over commodities. And it is also about the authority to make, enforce, or alter rules. The language used to debate what should continue to be done, and what should be done differently, is also a big part of stories about how commodities are ordered. To better understand global commodity governance today, vast insights can be gleaned from engagement with theories and approaches that are prominent in sociology, geography, international development studies, and political science. These include dependency theory, world-systems analysis, political ecology, and the political economy of development. If you are unfamiliar with any of those terms, a quick Google search or review of the literature could be instructive. But prior knowledge of those theories is not required to understand this book. As I indicated above, I do not belabor theory. I implicitly draw upon aspects of the first three approaches below, and present an overall analysis firmly grounded in the latter: the political economy of development.

Turning to cotton, global governance, and political change, globalization has contributed to opening up the spaces where the cotton order is created and performed to new players. And by globalization, I do not mean liberalization, privatization, deregulation or other phenomena associated with now out-of-fashion market fundamentalism. Globalization is about much more than the controversial basket of neoliberal economic policies that became known as the Washington Consensus or the "golden straitjacket." I should also note that globalization should not be understood to be the same thing as neo-imperialism or neocolonialism. In my view, globalization is a term that should be used primarily to denote the shrinking of time and space, and the implications of this faster, smaller world for people. Over the past decades, as trans-boundary connections between people connected to cotton have become more intense, and also more extensive, a faster and smaller world has yielded a very different cotton order. Where once a standard alphabet soup of states and inter-state international organizations held court, today's governance milieu features more diverse and fresher ingredients.

In seeking to set out and detail the politics of this thicker, spicier order for cotton, I should offer up two additional caveats. The first is that, as a political economist, my primary expertise falls in the area of development in the context of Africa. As such, I have tended in past publications to emphasize the implications of control over this crop for poorer and more marginalized people. And I continue to do so below. I make this choice not simply to emphasize the ongoing linkages between cotton and poverty, but to draw attention to the prospects for cotton to advance development. My view on what exactly development entails is very process-oriented, and is partly inspired by Amartya Sen.[9] For me, development is a qualitative process of social, political, and economic change. It does not start or stop when goals, targets, or indicators are endorsed or found to have been met. And it is fundamentally about enabling people everywhere to lead lives that they value more. You will hear this perspective come ringing through as we encounter situations where cotton has been controlled in ways that have prevented people from leading better lives.

Turning to the second caveat, I am fundamentally agnostic when it comes to the utility of cotton for development and the quest for sustainability. Put another way, I am a disinterested political analyst. I have no vested interests in this crop or in ways of producing or trading it, and I am not unidirectionally for or against cotton. As an analyst, I have searched for context-specific and contingent answers to the contributions that the exercise of control over cotton makes or does not make to sustainable development. Despite the United Nations endorsement of the sustainable development goals, this term remains an essentially contested concept. The Brundtland Commission's definition – meeting the needs of the present without compromising the ability of future generations to meet their own needs – was clear enough. But many assessments of sustainable development and cotton have continued to be highly subjective. Sometimes they have even been penned by those with vested interests who are much less open about their biases than I am. As an essentially contested concept, other genuine analysts and central players in the political dramas that I detail might advance understandings or assessments of sustainable development that are different than my own. And there simply is no way of arriving at a final, universal answer regarding whose definition or analysis is most appropriate. Bearing this limitation in mind, I do my best in this book to identify the consequential ideological flash points on development and sustainability that persist in the global governance of cotton. And from an analytical perspective and not from an ideological soapbox, I am convinced that numerous stakeholders in cotton have an interest in obscuring the politics of sustainability.

To launch the geopolitical analysis at the core of this book, I present my framework in chapter 2. This extended chapter introduces the prominent ideas, consequential institutions, and power relations that make up the world order for cotton. In it, I spell out my understanding of the evolution and dynamics of the really big-picture geopolitics. As this is the most theoretically inclined presentation in the book, it necessarily contains some generalities and a dash of abstract language. But fear not. The alarming realities presented in that chapter are scarier than the academic jargon. Thereafter, each subsequent chapter focuses on specific dimensions of the order and the related geopolitics. And to be clear, in those chapters I aim to speak to the politics of cotton directly. As such, I do not dwell on or retell the old chestnuts or the usual stories that experts might expect to find in a work on this subject. For me, as the teller of this politically oriented tale, it makes little sense to re-present the well-known social, economic, and environmental histories of cotton. Brilliant books on those painful histories have recently hit the shelves.

Consequently, in chapter 3 we turn to the politics of cotton on the land. To reveal this politics and make it more comprehensible, I have drawn heavily upon heuristics. As such, I offer numerous stories and examples that have been designed to emphasize a broad spectrum of land politics. These examples and stories offer models or approximations of the decision-making realities that cotton farmers navigate. In particular, they showcase a range of household land use, investment decisions, and trade-offs, and the associated politics. And the rationale that fueled this strategy is straightforward: perfect knowledge about political reality is not possible, given the scope and aims of this book and of the Resources Series. I could have offered a suite of anecdotal evidence drawn from across the world to back each and every one of my claims about politics. But the volume would have been at least four times as thick and much more of a chore to cut into. Instead, I have made a conscious decision as an educator to embrace a storytelling approach. It simply emphasizes the politics more fully and accessibly than any available alternative.

In chapter 4, the narrative turns to the politics of securing cotton for the country. Here I tell tales about the control that states exercise over minds and bodies. States generally seek to reproduce cotton to advance the "national interests" that they define. At times, those interests have taken the form of top-down efforts to secure more reliable or predictable foreign exchange earnings. At others, they have sought to make cotton work for industrialization. And at both times the social and ecological costs have been considerable.

I then proceed in chapter 5 to detail the politics of cotton when it falls into company hands. To do so, that chapter elaborates a range of scandals linked to the financialized fiber and details the emerging political contest over standards that aim to make cotton companies more "responsible" and "sustainable." I assess the legitimacy of financial cotton and physical cotton, and argue that there is a considerable risk that financialized business-as-usual might soon become cloaked in the language of "better" cotton. In chapter 6, I briefly offer my concluding perspective on what all of the geopolitics covered in the book means for sustainable development and for change in the world cotton order. Finally, after the conclusion, I present a humble anthropological account of my situation in relation to this fiber in the Afterword. This reflection or personal tale imparts a final piece of the geopolitical puzzle: the political orientation and engagements of this particular analyst with cotton and the world cotton order.

CHAPTER TWO

The World Cotton (Dis)Order

Imagine stepping into your favorite room. It doesn't matter where that room might be or what purpose that space might normally serve. Just think about the contours of that room and the normal order of things. And now imagine that room filled up to your waist and the waists of those around you with freshly picked cotton.

Confronted with such a scenario, most people would not hesitate to jump, dive, or otherwise belly-flop into the cottony softness. Some others might overthink the situation and question why their favorite room had come to be covered in cotton in the first place. But many would certainly temporarily suspend disbelief and immerse themselves in a moment of unthinking enjoyment. The out-of-context white stuff would ultimately be a source of passing pleasure, a quick escape from every day normality.

It is hard to envisage how violence could enter into this escapist tableau. Cotton bolls are not exactly the most effective potential projectiles. Security service personnel versed in "advanced" interrogation techniques, for their part, would assuredly attest that all of the raw cotton in that room would have little potential violent application beyond the threat of forced ingestion.

Stepping out of our imagined rooms and back into reality, warmongers and their historians have tended to treat this commodity as a secondary consideration. In this light, cotton has only mattered to world political history inasmuch as it has been needed to clothe the troops or "the masses." As such, in presentations of the vicious geopolitical conflicts that have animated the world disorder down to the present, cotton has tended to be portrayed as a bit player. It has been depicted as the quintessential docile or passive commodity. Cotton has been "lying in wait" for low-level functionaries or lesser leaders to defend, command, procure, and transform into apparel, bedding, or towels. Violence and cotton, from this point of view, has simply not been "page one" or bestseller material.

Suffice it to say that this conventional wisdom is as straightforward as it is wrongheaded. There is of course a grain of truth in the notion that human efforts to control cotton have stemmed from our need to defend ourselves against the elements. And – truth told again – it must be noted that cotton has only delivered marginal, fleeting victories over the variety of ferocious weather events that the earth throws at us. Getting our heads around the reality that efforts to secure our clothes have tended to go hand in hand with violence requires a willingness to question authority at all levels. Like verifiable information on what really goes on between the sheets, intelligence on the power and violence necessary to transform cotton into materials we can put on our beds has been patchy at best. The quest for a deeper understanding ultimately demands that students recognize that their cotton T-shirts are artifacts of world historical significance. The raw materials combined to make up these modest garments are enveloped by a geopolitics that is anything but simple.

But the good news is that the tools that permit weightier knowledge of this multifaceted twenty-first-century political economy are straightforward. The intellectual framework or approach necessary to understand the many dimensions of this complex politics comes from the field of international or global political economy (IPE). Many leading contributors to this emerging field have penned other volumes in the Resources Series. Policy and decision makers the world over increasingly seek out the perspectives of these scholars on geopolitics. The newfound influence of IPE is partly related to growing public awareness of the limits of purely economic perspectives on foreign and international affairs. To provide broader or more comprehensive views on resources in the global economy than economists that choose to work with abstract theoretical models, IPE scholars tend to draw on history. Their day-to-day research focus on what has been and is being done to people around the world by other people helps them to provide robust explanations of the politics of the present. In weaving their tales about how power is exercised today, they also foreground the importance of economic ideas and international and transnational economic institutions. Consequently, their stories about resource politics cover the United Nations and regional international organizations. They write about the activities of civil society, nongovernmental, and philanthropic organizations, and also about the deeds of "responsible" and "irresponsible" global corporations. Moreover, they cover what "developed," "emerging," "developing," and "least developed" countries get up to internationally and domestically. And, to top it off, they often draw attention to the powerful ideas, expert knowledge, and dogmas that animate political and business elites and infuse the transnational networks and coalitions that these non-state actors build.

So when it comes to explaining the geopolitics of cotton, sustained attention to mapping this vast sea of institutions, ideas, and power is required.[1] So too is a rethink of that safe, fun room of white cottony softness we imagined above. The nasty history of the violence committed to secure cotton over the years puts paid to the softness myth. The "fabric of our lives" has been covered with blood for centuries. Yet the myth persists. Scholars have written volumes on how the Atlantic slave trade effectively subsidized the development of industrial capitalism in the West. But when the blood flew and spattered the cotton plants on a fictional slave plantation in Quentin Tarantino's film *Django Unchained*, the symbolism was lost on many viewers. Cotton is a relatively hard crop to grow and has more often than not been associated with harsh conditions of life for those that rely on it. And to this day, international traders insist on labeling cotton – along with cocoa, coffee, and sugar – a "soft" commodity. Cleary, of the so-called "softs," cotton would be the hardest to swallow.

Economists that preach the benefits of free trade tend to laud the free cross-border movement of goods such as deseeded cotton that has not been spattered with actual human blood. They also tend to deplore the barriers that governments throw up to the international movement of cotton. And this is all well and good: trade barriers do generate social inefficiencies for some people connected to cotton some of the time. But on most geopolitical issues that IPE scholars would choose to foreground connected to trade, free traders seemingly have had cotton stuffed in their ears. For instance, early in its Industrial Revolution, Britain banned finished textile imports from India to enable its spinning and weaving industries to catch up to superior Indian production techniques. Along the same lines, many corners of Africa under imperial control and then direct and indirect colonial rule in the nineteenth century were sources of artificially cheap cotton exports for Europe's industrializing powers. The very idea that international trade in cotton has ever been "free" is itself a non-starter: it has always been subject to management, regulation, and control. And this goes doubly for industries that add value to cotton. Many states used cotton, trade barriers, and the public purse to build industry, generate employment, and clothe their populations.[2] Even today, companies, governments, and civil society groups actively attempt to control supply and demand for cotton in ways that expose what Cambridge economist Ha-Joon Chang has labeled the "myth of free trade."

This chapter commences by laying out the core organizational components of a framework for understanding the geopolitics of cotton. This approach aims to dispense with and ultimately situate the politics of free traders and of others of all political orientations with direct stakes in the maintenance of cotton's very problematic status quo. It deliberately stands back from and identifies political contests and debates in order to map the political forces and factors that bear upon development and change in the world cotton order. As such, the chapter spells out the roots of the geopolitical analysis that informs the rest of this book. Below, to start, the idea that there is a "world order" for cotton is discussed and developed. Thereafter, the components of this order are elaborated. To do so, the contributions that states make to the cotton order are articulated, and then the non-state actors that animate the order are detailed.

Thinking about world orders

The idea that there are "world orders" shot to prominence during and in the immediate aftermath of the resource or commodity booms of the 1970s. From the mid-point of that decade it was no longer possible to claim that scholars working to understand the challenges associated with the international commodity trade and economic development were working on a topic of marginal significance to international relations or international organizations.[3] After the oil-price spike of 1973, commentators in weekly economic and political news magazines started to refer to commodities as "high-level" world politics. Their rationale for doing so was sound. Developing countries had begun to work together to advance their shared interest in securing better prices for their commodity exports. European imperialists and colonialists had previously imposed the economic structures and activities in most territories where new countries sought to advance their interests. Unsurprisingly, the dominant structures and activities in developing countries were linked to the extraction or production of primary or raw materials for export.[4]

Given these broadly shared limiting conditions, the supply control efforts of the Organization of the Petroleum Exporting Countries (OPEC) were only one of many important dimensions of the newfound collaboration amongst developing countries. The quest to make resources pay and achieve greater autonomy from old colonial trade patterns increasingly informed the work of the international movements and groups that brought these countries together. The Non-Aligned Movement of developing countries that endeavored to stand apart from the Cold War blocs of "West" and "East," and the Group of 77 developing countries at the United Nations, became vehicles to advance the resource liberation agenda. Vijay Prashad in his book *The Darker Nations* subsequently termed this quest the "Third World project" to make the world economy work better for development. In the wake of this networking, industrialized economies responded with their own collective efforts, forming the International Energy Agency (IEA) and pushing back at the United Nations. The so-called "North–South" conflict over the world order was born.

Surveying this scene, Robert W. Cox, one of the leading intellectual pioneers in the field of international political economy, identified the contours of the postwar "liberal international economic order."[5] He juxtaposed this western- or northern-dominated order with the "New International Economic Order" (NIEO) that developing countries of the Global South called for within and beyond the United Nations. The former, in his eyes, was centered in the North Atlantic and advanced progressively liberal international trade and the roll-out of social policies associated with the welfare state. In contrast, the NIEO was a comprehensive package of prescriptions for a new order that aimed to enable sustained economic development outside of the core industrial countries, and it prominently featured a detailed plan to better manage the commodity trade. Given the standoff between the North and the South over the order, Cox's scholarship engaged with the term "order" and its use.[6] His work on this topic pushed academics and policy makers to consider international relations to be about more than what states do individually to advance their own interests. He also encouraged analysts to look beyond the liberal notion of also paying attention to what states do collectively outside of their own frontiers to cooperate with each other and advance mutual interests.

In a nutshell, Cox argued that these traditional aspects of international relations scholarship mattered, but they could not be understood to offer a full picture of the politics of "ordering" beyond borders. For Cox, the operations and power of bricks and mortar inter-state, international organizations, and the transnational activities of firms involved in the production and trade of resources and other goods were worthy of equal attention. Moreover, his work, rooted as it was in the context of the divisive North–South debate, underscored the reality that conflict over ideas about fairness and justice mattered at a global scale.[7] Power in the world order, from this point of view, was about much more than what individual states wanted to do with their guns. It was also about what they and others wanted to do or not do to manage "butter," or more accurately, resources. In this light, power was deemed to be about the capacity that states, international organizations, businesses, and other transnational groups commanded to advance ideas about global economic stability, stasis, or change.

In a curious twist of history, the politics of world ordering that gave rise to Cox's intellectual innovations met a slow death as his publications splashed into the literature. More accurately, this politics was actively "disappeared." Progress on the NIEO was frustrated after the Third World movement confronted a brick wall of international talk shop meetings and fora. This negotiating approach had been agreed at the United Nations in 1975 on the grounds that it could break the North–South impasse. It ostensibly aimed to facilitate progress on the South's demands for a new order through fostering discussions or negotiations linked specifically to individual components of the NIEO. However, this strategy ultimately broke the NIEO. Looking back, it devolved into a "divide and rule" situation. The North's power was left untouched as negotiations ground on, and the South was left more divided and effectively ruled. The profusion of issue-specific negotiations that ensued after 1975 simply outstripped the capacity of the South to collaborate effectively. And, as this capacity gap grew, the priorities of governments that had initially supported the NIEO package diverged. Some supportive voices were lost as governments changed and, as varying degrees of export-led success were realized across the Global South, this resource-linked politics faltered. And then it ultimately fell off the rails when commodity prices tanked, interest rates in the West skyrocketed, and Margaret Thatcher and Ronald Reagan challenged the global Keynesian consensus that had underpinned the old liberal international economic order.

The success of the West's push back against "the rest" only reinforced Cox's conclusions. Powerful ideas and institutions were implicated in political conflict beyond borders to order the world. And as the world changed, so too did the order. Drawing theoretical conclusions from his empirical cases, Cox emphasized the necessity of mapping and detailing multiple levels of politics to understand transnational or global-level politics. He argued that world orders were configurations of political ideas, institutions, and "material capabilities." The latter, in the context of the twenty-first century, can be termed "power." But the complete picture of world ordering did not end there for Cox. In his view, due consideration needed also to be given to the states that formed part of the order, and also to the ideas, institutions, and power dynamics that made up each state in the order. Furthermore, he argued that it was essential to detail the ideas, institutions, and power dynamics associated with "social forces" or non-state actors that operate across borders, such as businesses and the labor movement and their associated networks and coalitions. Today, in addition to these prominent transnational non-state actors, other groups that are configurations of influential ideas, institutions, and power now warrant attention. These non-state actors include the non-profit, nongovernmental groups that compose global civil society, and also the group of 200,000 ultra-high-net-worth individuals that command a net worth of at least US$30 million each.

What all of the abstract theoretical language above means for the world order, and for cotton, is thankfully quite simple. The categories Cox employed when he argued that "states, social forces and world orders" are demonstrably configurations of ideas, institutions, and power, offer simple analytical tools through which the geopolitics of cotton can be framed and elaborated. There is of course a world order that is much broader and more encompassing than the order connected to cotton. But the aspects of the world order that intersect with cotton can be rigorously identified and situated using Cox's categories for explaining the broader geopolitics that encompasses humanity and all resources globally. In a sense, then, like the other commodities covered in the Resources Series, this book employs a few of the tools IPE scholars have deemed necessary to produce robust explanations of world ordering. It does so to elaborate the situation of one seemingly innocuous commodity in that order that we all see or use every day.

Much like the world orders that seek to govern other resources of global-level significance, the order for cotton has become increasingly complex. In the 1970s and into the 1980s, most governance contests over cotton primarily played out at the inter-state or international level. Debates between producing countries and consuming countries were where the political rubber hit the governance road. State-to-state contest over production and marketing strategies, prices, trade volumes, quality, and other issues was a time-consuming reality. And an international organization that brought together producing and consuming countries – the International Cotton Advisory Committee (ICAC) – was front and center in numerous debates over cotton exports and imports. Over those years, cotton also faced growing competition from synthetic fibers derived from petrochemicals. This particular marketing challenge made cotton a high-level concern for many big exporting countries. It worked against the agenda of exporters in the

Global South that continued to collectively push for a more remunerative cotton trade. The polyester threat was very real, and it divided the developing world where cotton was grown and also sold abroad.

So when in the context of the broader movement to make commodities work better for development, the United Nations Development Programme (UNDP) sought to build a new international arrangement for cotton, the international politics was fraught. However, as the NIEO agenda faltered, so too did progress on this proposed international commodity agreement (ICA) and organization known as the Cotton Development International. Little headway was made to establish a global organization that would manage the global supply of cotton in the interest of economic development. While international commodity agreements for other "softs," such as coffee and cocoa, were agreed and held up into the 1980s and beyond, the hard reality for cotton is that no such agreement ever saw the light of day.

One international agreement nonetheless weighed heavily upon the successes that could be reaped from this crop. International trade in textiles and garments remained heavily regulated during the postwar liberal international economic order and became increasingly so from the mid-1970s. Advanced industrialized states that had employed their public resources to develop private industry in this area did not seek to bring textiles and garments into international trade liberalization negotiations. The push by members of the General Agreement on Tariffs and Trade (GATT) – the precursor to the WTO – to liberalize international trade did not apply to textiles or garments derived from cotton, polyester, or other fibers. Rather, from 1974, export quotas for products made from these fibers were allocated under the Multi-Fibre Arrangement (MFA). The MFA was to be a temporary measure to enable countries that had already developed significant industries to stem the tide of simple textile and garment imports from the Global South, and transition their economies away from those manufacturing activities. It lasted for thirty years, and, according to the International Monetary Fund (IMF), came at a huge financial cost to the developing world.

The MFA reality put a serious upward constraint on the capacity of cotton exporters in the Global South to make the choice to invest in new industries that would enable them to export more than raw cotton. While the arrangement held, any progress developing countries made to diversify their cotton exports was consequently frustrated. Cotton products that had had more done to them than the simple removal of seeds, through a process known as ginning, and subsequent packing into bales for export were subject to strict quota limits. While these limits were negotiated, like other negotiations in that era, developing countries exercised less voice or control than advanced industrialized countries. The arrangement itself was designed to be a constraint on their capacity to use cotton and other fibers to pursue economic development. As such, it was a risky proposition for developing countries to make investments that would enable their cotton to be spun into thread and then weaved into fabric for piecing together as garments for export. In many cases, such efforts were ruled out a priori. Where and when they were realized, they either foundered or were by and large restricted to attempts to serve smaller or more limited domestic and sub-regional markets. To be clear, the "inefficiencies" of industries linked to cotton across the developing world did not derive from the unduly "interventionist" efforts of misguided policy and decision makers. Rather, state-supported cotton industries were held back as the direct result of a trading arrangement that actively favored those in Europe and North America that had already developed and that had also used the state to do so.

In sum, the idea of trade liberalization was simply not put into practice in the world order for cotton as free trade and market-liberal policies became more politically resonant in the 1980s and into the 1990s. Governance contests were primarily about how to best go about fending off the synthetics onslaught. And they were also about how quickly developed countries would be asked to adjust to the reality of export-oriented industrial development in the Global South. Almost down to the end of the twentieth century, ideas about the implications of cotton farming for the health of those that grew it or for the land upon which it was grown were simply not considerations of global-level significance. The institutions that mattered for production were primarily domestic in origin. Corporations that bought or sold cotton across borders also engaged in governance. But beyond the analysis of fibers at ports of entry or exchanges, cotton traders and quality professionals were secondary players in the order. States and the international organization designed to bring them together by and large reigned supreme.

The new cotton order

And then something radically exciting happened. As the internet enabled farmers, their advocates, and socially and ecologically concerned groups to increasingly network across borders and raise awareness about challenges linked to cotton, new ideas and new institutions started to have an impact on the world cotton order. Prominently, in the mid-1990s, students in the United States and members of the global labor movement documented and trumpeted the horrendous working conditions linked to many textile and garment industry operations in Southeast Asia and elsewhere. Debates over cotton and the goods derived from it were no longer dominated solely by the states that had an interest in producing the crop, trading it, or reaping industrial development from it. In the wake of this newfound attention, governments and corporations reconsidered their positions in relation to the status quo of cotton. Some textile and garment companies even started to talk about "responsibility." Clothing retailers and cotton traders soon after realized that there were considerable risks associated with business-as-usual. Environmental activists and human rights campaigners pushed topics as diverse as the salinization of fresh water and child labor into the space where ministers and their international delegates had formerly exercised the unchallenged right to govern cotton. While new cotton powers such as China and Brazil emerged and plowed more resources into industrial-scale production, other innovators working at smaller scales developed ecological approaches to growing cotton. The latter have fundamentally challenged the notions of quality, productivity, and efficiency that informed twentieth-century approaches to industrial cotton farming. Taken together, these dynamics and others detailed in subsequent chapters radically altered the status quo and contributed to the emergence of a new order to govern cotton.

In essence, the world order for cotton in the twenty-first century is composed of consuming and producing states and also of non-state actors including businesses and civil society groups.[8] The politics of this order at its most abstract is relatively straightforward. States and non-state actors connected to cotton draw on divergent ideas about cotton production and trade to advance their interests. They also seek to inform, build, and influence institutions that bear upon the prospects for cotton to contribute to sustainable global development. As such, they wield power. And it is important to be specific about what exactly "power" means to understand the power dynamics of the new order for cotton. To do so, the perspective advanced by Jennifer Clapp and Doris Fuchs in their book *Corporate Power and Global Agrifood Governance* is highly instructive.[9] Building upon earlier power analyses in the field of IPE, Clapp and Fuchs identify three dimensions of power. In their estimation, when these three dimensions are fully considered, the spectrum of global power politics related to the exercise of control over agricultural commodities is revealed.

The first dimension of power can be referred to as *direct* or *relational* power. Many people tend to think about this aspect when they first envision what the exercise of power looks like. This dimension is all about the power to get others to do something that they would not otherwise do, for example, the power exercised via the barrel of a gun or the provision of a payment to stop an invasion or release a hostage. In the area of resource geopolitics, this type of power typically takes the form of lobbying at the domestic and international levels. It also tends to be tied to the financial resources that particular states, businesses, or civil society groups command and have at their discretion to directly influence the perspectives or actions of others. One prominent historical instance of the exercise of direct power in cotton must not be forgotten: the power that plantation owners and slave traders exercised over enslaved people who would not have otherwise tended to this thirsty, weed-prone, and difficult-to-harvest crop.

A second dimension of power can be found in the power that governments, corporations, and non-profit groups command to set the rules of the game. This aspect is consequently about the power to organize or control the generally agreed framework that shapes and constrains the actions of all players. Some IPE scholars refer to this dimension of power as *structural* power or "rule-making authority." Regardless of the terminology, the power to structure the actions others take has been a big feature of IPE scholarship on the history of global commodities. Sticking with the example of cotton and slavery for a moment, during the imperial and colonial eras, European states effectively set the rules that permitted the transnational slave trade. The Abolitionists, working domestically and also across borders, subsequently attempted to countervail this trade. To bring an end to the direct and awful power that slavers and their clients exercised over so many lives, they challenged the states that set the bloody rules of that filthy game. And in so doing they also contributed to laying the foundations for the emergence of a new rule for large-scale cotton production: workers had to be free, willing, and paid.

The final dimension of power that political actors employ globally relates to what they actually articulate or say, and to how they articulate or say it. IPE scholars tend to refer to this aspect of control or domination as *discursive* or *linguistic* power. A popular expression that captures a piece of this puzzle is "jawboning": the power to persuade or exert pressure through language. Authorities tend to resort to jawboning when they are unwilling or unable to pull the trigger, disburse funds, or make new rules to advance their interests. But the exercise of linguistic or discursive power does not start or stop there. Actors with little to no capacity to exert direct or structural influence can and have used language effectively to reshape business-as-usual. Returning to our Abolitionist example, first as isolated writers and subsequently as contributors to a loosely organized movement, the Abolitionists reframed the language that politicians and industrialists used to describe slavery. They challenged the notion that people could be turned into simple relatively undifferentiated commodities to be bought and sold to serve the needs of industry. And they introduced questions of morality, justice, and ethics that had previously been absent from sites of slave production, trade, and consumption. As such, the Abolitionists demonstrated that the power of persuasion often depends on reframing the terms of political discourse. The old discourse had been about the functioning of the industry. For instance, the concerns that "legitimate" slave traders articulated regarding the eradication of piracy and the realization of greater maritime "security" were front and center in many political discussions. Through using new language to frame the rules of the game as illegitimate, the Abolitionists changed the order for cotton, and also the world.

The frustratingly complex reality of power is that none of its three primary faces ever appears in isolation. There are no guarantees which aspect of power will dominate where ideas about cotton are contested, and when institutions that govern cotton are built, renewed, or transformed. But when the exercise of power is viewed through Cox's simple categories of world ordering, it is possible to produce a coherent, simplified account of the grand contests that shape political stasis and change. His framework enables us to stand back from this power politics in order to situate it and help people involved in it or subject to it to better understand it. In this light, the power politics of cotton in the global economy is about much more than farmer subsidies and world prices. As the eminent political economist Susan Strange might have put it, it is about who actually gets what from cotton, where, and when.[10] Put simply, the transnational, multi-level power politics is much broader and deeper than most businesses, governments, and international organizations have as yet been willing to publicly admit.

States in the cotton order

Sovereign territories that engage with cotton come in many different varieties. States that produce, export, or import cotton do not do so in the same ways or in the same volumes. Cotton is to many a marginal part of their broader political and economic activity. But for others, cotton provides their lifeblood: their principal source of export earnings denominated in hard currency. Flows of cold, hard US greenbacks are not the end of this story, however. Some states see cotton as the ideal crop that can be used to incubate and nurture new or experimental agricultural technologies, such as tractor satellite systems or genetically modified seeds. States that aim to generate efficiencies through further industrializing the sector, and that rely on global agribusinesses to scale up production, are nonetheless in the minority. Others primarily depend on the labor of farmers and their families themselves. Hand sowing, hand hoeing, hand-picking and the packing of donkey carts by hand gets the job of cotton done in many diverse places across Africa and Asia in 2016. And even in one of the great hubs of industrial agriculture at the core of the cotton economy – the United States – alternative, non-industrial approaches to cotton have taken root. Looking beyond borders to the order where states seek to advance their own interests internationally, they often do so independently or on their own to challenge other states. As such, it makes little sense to cover the priorities of states in this order with broad brushstrokes.

That being said, it is possible to identify the contributions that states make individually and collectively to advancing, perpetuating, or challenging particular ideas about cotton. For instance, many states maintain policies that aim to bolster and transform their cotton sectors. Agriculture policies in the United States and in Europe have offered strong support to cotton producers to help them stick with this difficult crop. These supports have typically come in the form of various grants, subsidies, or credits. And in the case of the United States, certain benefits that large-scale cotton operations have received have contributed to increasing the volume of cotton produced worldwide. Since US-based manufacturers are no longer major global players in the business of transforming cotton into textiles and garments, US cotton support policies have pushed US exports of raw cotton higher. This reality had a significant long-term negative impact on the world cotton price well into the 2000s. As such, it undercut the interests of cotton-exporting countries that lacked the financial resources to furnish their producers with similarly sweet deals. Several emerging and developing countries, including Brazil and a grouping of African countries that depend on cotton known as the "Cotton Four" (C4) – Benin, Burkina Faso, Chad, and Mali – have continued to push back against this status quo.[11]

The idea that states should facilitate, respond to, and resolve the recurring events and crises that work against cotton and the windfalls that can be reaped from it has remained broadly shared. Cotton-producing states the world over continue to step in to facilitate resolutions to periodic crises related to spikes or troughs in the world price. They provide compensation to farmers when their lands are afflicted by droughts or when crops fail due to seed nonperformance, blight, or pest infestations. In recent years, many states have even bailed out firms or taken direct control of troubled operations. When businesses that provide producers with seeds, fertilizers, or pesticides have failed, or when the enterprises that buy cotton, remove the seeds, or market it internationally have gone bankrupt, states have stepped in. But, to be clear, they have not done so to the extent that they once did. Over the past twenty years, many state-run firms that once held monopolies over the provision of inputs to cotton farmers, or that were the only licensed buyers of cotton in particular countries, have been broken up and privatized. Competition amongst private firms is something that many states now endorse. Yet states continue to make essential contributions to the performance of cotton every day. Through their ongoing domestic policies and crisis responses, states in the world cotton order effectively endorse the idea of active government management.

However, the same cannot be said for the idea that the global cotton supply should be subjected to inter-state, international management. That idea was front and center in the international community of official policy and decision makers, diplomats, and negotiators after OPEC leveled the "oil weapon." At that time, governments took the international politics of the business of commodities as diverse as rubber and tin very seriously. And they did so with good reason. According to the then influential economic analyses of the world's self-proclaimed "number one" free trader, Jagdish Bhagwati, the over-production of raw materials by export commodity specialists can fuel growth that is "immiserizing" or impoverishing. In the context of global production increases, his point was that exporters would have to sell more and more year on year to realize consistent returns. The controversial research of Raul Prebisch and Hans Singer on resources and commodities went even further and was even more prominent internationally. Working independently, Prebisch and Singer noted that the terms upon which commodity-exporting countries traded were subject to perpetual decline. What this meant was that, over time, resource economies had to sell more commodities than they had done previously to be able to purchase and import the same amount of finished goods. In the case of cotton, for example, if a given country wanted to import ten state-of-the-art combines or tractors per year, they would need to sell more cotton every year than they had the year before in order to continue to do so. Given this analysis, many high-level officials concluded that the commodity trade generally, and the cotton trade specifically, needed to be actively managed to facilitate economic development.

In their quests for supply management, many governments attempted to apply internationally some of the thinking that had fueled their own domestic agricultural supply policies. International discussions between commodity exporters and importers homed in on the need to manage prices so that they stayed within ranges or "price bands" that were politically acceptable for sellers and buyers. Those that were tasked with negotiating new commodity agreements also drew upon their own national experiences of physically compiling stocks of commodities. Many governments maintained stocks of key food and non-food commodities to weather domestic price or supply shocks. If a maize crop failed and imports did not fill the gap, for instance, stocks of corn could be released to lower market prices. In the interest of supporting its farmers, as Europe developed a common agricultural policy, it rolled out the world's most elaborate stocking network for commodities including butter and even wine. So it should come as no surprise that perspectives on the need to establish international stocks informed the North–South debate and international commodity politics as it reached a high-water mark. Supply management was a topic literally as European as aged cheese.

A buffer stock for cotton subject to active international supply management and price control in the interest of producers, consumers, and development nevertheless failed to materialize. While support for such an initiative emanated from governments across Africa, Asia, and Latin America into the 1980s, today states generally hold the idea that the world price should be set by market conditions. This idea is widely preached at international gatherings on cotton. But, as indicated above, it departs from the reality that the domestic policies and crisis responses that states enact have considerable and consistent impacts on the volume of cotton traded internationally. While states generally defer to market participants when it comes to the value of cotton – its world price – if they ascertain that actions taken by other states domestically to secure production volumes have undercut that price, they tend to cry foul and argue for "free trade." The idea of free trade is thereby consequential internationally. It can be used to call out real or imagined price "manipulators." And the efforts which states make to call time on the practices of other states they deem to be too price manipulating are a far cry from the old inter-state efforts to control prices. Free trade is now a rhetorical tool. States can utilize free-trade language to blow the whistle on other states that they deem to have been too heavy-handed in their domestic supply management schemes. Yet all states in the world cotton order remain implicated in such efforts now that international supply management has been abandoned. In its place is the new idea of effective price "management": the real or potential threat that a state might be disciplined for being relatively too lavish when it comes to cotton and cotton-producer support.

States also advance other ideas about cotton that matter internationally. Governments that produce and trade this commodity implicitly endorse the notion that efforts to produce cotton are better than not producing or exporting cotton at all. They do so by not playing up alternative crops or fibers too much on the international stage and also through explicitly preaching the virtues of cotton. And there are certainly many viable alternatives to conventional cotton that could feed and clothe more people and do so more durably. But the systems that states have developed and maintained to build the capacity of their cotton sectors represent financial and human investments of historic scale. In all countries that produce cotton, there are now masses of people that command skills specific to this commodity business that are not necessarily transferable to other sectors. As such, cotton is in part about the paths that states have walked with it and their ongoing dependence on the paths they have chosen. As cotton systems continue to function and markets for cotton continue to exist, states have an interest in portraying it to be anything but a sunset industry. While pieced together into a garment, cotton will always perform poorly versus the elements relative to other natural and man-made fibers, states contribute to keeping us hooked on "the look" and "the feel" of cotton.

Finally, states also contribute to propagating the newer idea that businesses involved with cotton have a responsibility to redress social and environmental challenges. Many states have made the choice not to create extensive rules or legislation to promote more sustainable or remunerative alternative farming approaches domestically or internationally. As such, most have ceded this terrain to business self-regulation initiatives and so-called "best practices" and standards that bolster corporate public relations.[12] Through actively or tacitly delimiting the scale and scope of their own governance initiatives for cotton, many states have endorsed the new conventional wisdom that ever greater reliance on corporate social responsibility (CSR) can deliver social and environmental "wins." The idea that companies can do good for cotton and those that farm it is not necessarily untrue. But the prominence of CSR must be understood to be associated with a considerable opportunity cost: aborted or forgone universal public initiatives that could afford all cotton farmers better access to demonstrably more responsible practices. In its wake, distinct tiers of producers have emerged. There are those that can access various kinds of CSR systems. And then there are those that remain subject to suppliers and buyers who adhere to free-market economist Milton Friedman's infamous dictum that "the business of business is business." It must also be emphasized that life for those in the former grouping is not all about wine and roses. Not all CSR approaches deliver the same extent of responsibility or remuneration. As will be elaborated later, light-touch and low-cost "types" of CSR command considerable power over the livelihoods that families can scratch from cotton.

Turning to the state-based institutions that make key contributions to governing the world cotton order, trade and foreign ministries are front and center. Delegates from these ministries to the WTO and to other international and regional international organizations matter. Representatives to the African Union, for example, have led discussions on the proposed multi-stakeholder Pan-African "road map" for the development of cotton. To reiterate, this EU-supported road map aims to enable African countries to work collectively to add more value to cotton and better serve local, regional, and global markets.

For their part, agriculture ministries and cotton marketing boards also tend to have impacts on the order. The policies that they implement with the aim of supporting their own cotton sectors are certainly domestic in orientation. But the activities they engage in domestically, and the choices that they make in support of particular interventions, can and do have considerable international spillovers. The actions or reports of the United States Department of Agriculture (USDA) and ministries across the European Union do not stand alone in this regard. For instance, when authorities in China made a decision to draw down that country's massive domestic cotton stockpile in late 2014, this new direction changed everything for Africa's cotton producers. The release of stocks reduced demand in Chinese mills for African cotton the following year. Cotton marketing boards and ministries on that continent were then left with an unenviable choice. They either had to inform their producers that they needed to plant more to make up for a presumed decline in the world price. Alternatively, they needed to get the word out that it was a good year for farmers to direct their focus away from cotton to other cash or food crops. How much hardship or hunger China's move ultimately generated in the remote zones where cotton is produced elsewhere is not yet known. But its impact on the world has been considerable. In this instance, the absence of evidence does not imply the evidence of absence. China's unilateral move assuredly exported pain.

Similarly, when Brazil decided to scale up cotton production through granting its biggest farmers subsidized access to the latest industrial farming technologies, the fallout was also global. Once many Brazilian tractors on massive monoculture operations were equipped with geographic positioning systems and other satellite imagery technologies, it was not long before production and then exports surged. And here, too, in the wake of Brazil's measures in places such as Mali, farmers who depended on flows of cash from cotton had tougher times keeping their impressive oxen teams and donkeys well fed. Not to mention their children.

In both large and small cotton-exporting countries, other government agencies have made big contributions to ordering cotton. Where and when national-investment promotion agencies have enticed agricultural investors with an interest in cotton, these state-backed efforts have altered the distribution and volume of production globally. Similarly, foreign investment and competition regulators have had big impacts on the growth and success of the commodity trading houses that deal in cotton. They have done so through exercising their responsibilities regarding foreign mergers and acquisitions. When Australia approved Singapore-based Olam's takeover offer for Queensland Cotton in 2009, for example, this approval enabled an upstart cotton trader to secure a stronger foothold in a global business long dominated by family-run firms. The following year, Louis Dreyfus Commodities cotton unit Allenberg Cotton received approval from the United States to acquire Dunavant, a long-standing global cotton trader. But the potential impact of domestic state-based institutions on the global order does not end there. In all cotton producing and exporting countries, some of the biggest state institutions that enable and constrain cotton are part of the architecture of the everyday life of business. The standards, weights, and measures agencies. Rail, road, and port authorities. And of course, the police, customs, and tax collection bodies. Taken together, these domestic institutions can facilitate the tradability of cotton. Conversely, they can also undermine individual players in the industry or even the viability of the sector itself.

The latter points on state institutions in the cotton order speak to the need to engage at a deeper level with an analysis of power. It is not sufficient to simply present the ideas and institutions through which states contribute to creating and maintaining the international political economy of cotton.

What states endorse, and what they do, can be seen to wield power relative to other states. We have seen that states can use the ideas and institutions at their disposal, such as the idea of free trade and institutions including the WTO, to countervail the power other states exercise over cotton. But what is said and done by states can also empower and disempower non-state actors, including businesses, individual farmers, and civil society groups. As we have seen, state actions that align with and advance business interests, such as the direct power that states command to provide incentives to large-scale agribusinesses, can harm farmers in other states or disempower smaller farmers at home. And when states relinquish power, this can also bestow power upon some organizations and individuals in the order, or take it away from others. For instance, when states cede responsibilities that they formerly considered to be part of the rules of the game for cotton to "the market," such as giving up on the idea of international price management, this can empower business. Those that trade physical cotton or that speculate in the numerous financial products whose value derives from cotton can be empowered. Moreover, when states articulate viewpoints on the economic necessity of cotton to the growth and development of their economies in spite of evidence to the contrary, the exercise of discursive power can easily disempower people. It can condemn farmers who buy into bogus narratives on the benefits of cotton to food-insecure, precarious futures.

These broad brushstrokes of the state-related dimensions of the geopolitics of cotton nevertheless offer at best only a partial picture of the global politics of cotton. Engaging with the ideas and institutions held and created by the prominent non-state actors that compose this industry beyond states – including big businesses and others that seek alternatives to business-as-usual – can help to complete the portrait. Non-state actors are intimately involved in the global governance of cotton or its world ordering. And it must be noted that farmers themselves fall across both of these categories. Some high-turnover industrial-scale producers and the organizations that advance their interests consistently fall on the business side of the ledger. But when smaller farmers work together with each other or with farmer research or advocacy groups and smaller businesses, the ideas and institutions they give voice to are often quite distinct from those that bigger producers support.

Big business and the cotton order

Transnational agribusiness corporations seek to advance ideas about the particular technologies that are most appropriate for cotton. In some cases, these firms reap direct benefits from their advocacy for particular technologies. Monsanto, for example, developed and holds a patent on cottonseeds that have been genetically modified or encoded with proteins from other sources that are toxic to insects. Through inserting a common soil bacterium (*Bacillus thuringiensis*) to transform these seeds, this firm has aimed to develop cotton that is more resilient to infestation by bollworms and other pests. And to recoup its research and development costs and profit from this innovation, Monsanto has endeavored to brand this new technology as farmer-friendly and sustainable. Since so-called Bt cotton should in theory reduce farmer use of agrochemical pesticides, there is potentially or in theory a grain of truth in the language this firm has used to package and sell this product. Monsanto's marketing efforts have had a big impact on the cotton order insofar as states as diverse as the United States, India, and South Africa have authorized Bt cotton and encouraged their farmers to embrace it. Whether or not the cotton systems in those states and others are the better for it is now a hotly contested topic in academic and policy circles.[13] The power of this patent and of patents generally over people and the planet is the stuff of serious cotton politics.

Other agribusiness firms that produce seeds, fertilizers, herbicides, and pesticides that have been tailor-made for cotton also produce reams of public information on the particular benefits of their products. Unsurprisingly, advertising and marketing materials for these agricultural "inputs" tend only to touch on product characteristics and impart directions for application. Firms push the idea that the appropriately timed and sequenced application of a suite of conventional agrochemicals is necessary to successfully maximize farmer yields of cotton elsewhere. In particular, they do so in the promotional materials that they produce for farmer field schools, agricultural colleges, and ministries, and also in farm supply magazines and trade publications. This hands-on, direct marketing approach has garnered adherents to the view that agrochemicals are essential to ensuring yields and quality. From Uzbekistan to Tanzania, many local businesses that buy cotton and assist farmers in their quests to produce it have bought into this corporate public relations.

And make no mistake: at one time, the intensive use of chemicals was the only sure bet for cotton. Farmers who bought into this package and reaped success produced record volumes of incredibly white fibers. There was some truth to the notion that this package was the road to producing the whitest, most sought-after crop. Today, however, some have painted a different reality than the PR story. These experts assert that the biggest pest management innovations over the past decades have not come from modified seeds or petrochemicals, and posit that artificially white, environmentally unfriendly cotton is past its use-by date.

The story of sales and marketing ideas in the world order does not end there. Big firms market implements, tools, tractors, and computers that are used by farmers and their buyers. Corporations with farm machines to sell tend to durably engineer ideas about why those devices should be considered farm essentials. While the invention of "needs" for farm technologies – so-called "demand creation" – is not exclusive to cotton, the development and sustainability challenges facing cotton warrant attention to the brokering of ideas about how it should be grown. When firms push a new farming technology that can replace implements or tools that are currently in use to do the same job, they become actors in farm politics. Where cotton is grown in remote regions of countries where most people continue to earn their incomes and livelihoods from farming, this politics is especially stark. The greater uptake of tractors equipped with satellite technologies in the south and east of Senegal, for example, might at first glance seem unidirectionally positive. More tractors and know-how associated with their use could in many cases help to assure that more cotton is produced more consistently by some farmers. Yet their introduction could also be associated with income losses and other harmful effects, such as the unsustainable compaction of relatively thin soils. The history of the modernization of agriculture is littered with examples where the introduction of technologies that are inappropriate to a country's level of development has yielded rural out-migration, joblessness, and landlessness. Control over ideas about what technologies are most appropriate for cotton in particular places is thereby of world historical importance. This is the ideological terrain upon which the expansion of industrial farming systems for cotton is contested and secured.

Even so, corporate ideas about cotton-production technologies are not the only business ideas of consequence. Of equal significance are the ideas that big businesses hold regarding the role of finance and of financial products. Commodity traders that peddle cotton and the financial institutions that provide finance to cotton producers and traders advance the idea that participants in the cotton market should ultimately set and manage the world price. From this point of view, the businesses that directly benefit from cotton are most qualified to "discover" and manage prices. These firms have an interest in assessing supply and demand conditions for cotton accurately. They have also rolled out products that market participants can use to manage risks associated with supply or demand shocks. On the former, risks posed by droughts, blights, the mass sale of old stocks, or a surge of cotton from new or newly high-tech producers can now be managed. The same goes for the latter shocks, such as recessions or consumer preference for other trendy fibers and fabrics, such as linen or hemp.

The new conventional wisdom on price is that private actors that engage with each other at the exchanges where cotton is traded or who "meet" in the electronic market command the most reliable price information available. As with other agricultural commodities, the spot price at which cotton can be bought and sold today is only part of the new price/risk-management story. Business has endorsed the view that contracts to sell or buy specified amounts of cotton in the future are acceptable risk-management tools. As well, they have pushed for and established slightly different futures contracts that give their bearers the option to buy or sell specified amounts of cotton. And to cement the idea of corporate control over prices, they have in many cases advocated successfully for market regulators to enable market participants that do not hold any physical cotton to engage in the buying and selling of futures contracts linked to cotton.

Whether any of these price/risk "innovations" have yielded more reliable price setting and management remains an open question. The counterfactual – an official inter-state, international system to stock cotton and manage its price – would be associated with significant costs. But so too is the current system. Perhaps inordinately so. Since the business of world price setting and risk management was effectively privatized, the volume of financial products whose value is derived from cotton has exploded relative to the actual growth of physical production volumes worldwide. Put another way, the application of financialized price setting and management – at least on paper or on computer terminals – has inflated the overall economic value of this industry. As such, the financial side of the cotton business now makes the industry appear to be much bigger than the "real" cotton economy.

And incentives and priorities at the apex of the merchant industry reflect this shift. For many businesses that generate returns from trading futures, options, and other complex financial products linked to cotton, developments in the production of physical cotton are now a distant or at best secondary consideration. In other words, in the new global financial life of cotton, the risk-management strategies of others with skin in this game take precedence in everyday decision making. Firms actively engage in speculative trading approaches that aim to beat the strategies they presume other firms to have adopted to manage their cotton assets. Such speculation moves markets as reliably as the marquee agricultural reports on physical production volumes issued by the USDA and other agencies. The ongoing salience of the corporate ideas that built the new status quo and contribute to reproducing it has deep implications for the people that grow cotton and for the possible uptake of more sustainable or equitable approaches to growing it.

Beyond the agribusinesses and cotton traders mentioned above and their principal competitors, including Cargill, Noble Agri, and Glencore, several other types of business institutions underpin the increasingly corporate world order for this commodity. Firms that move cotton within and across frontiers from seed-removing ginneries to thread-creating spinners are also key players in the cotton value chain. These include companies that provide trade finance or bridge loans, transport and logistics businesses, insurers that insure physical commodities and commodity-linked assets, and corporations that transform lint into thread and textiles. Another type of organization – business associations – seeks to inform the governance of the financial and physical cotton trade, and in some cases aims to advance business control. National associations of cotton-linked businesses make powerful contributions to the governance of this sector around the world. And other pan-industry groups, such as the International Swaps and Derivatives Association, advocate globally on behalf of firms that trade commodity-linked instruments in support of the lighter-touch regulation of financial products.

Transnational business networking also continues to matter. The International Cotton Association (ICA) was formed more than 175 years ago by a group of cotton brokers in Liverpool to facilitate the cotton trade. The ICA today boasts hundreds of corporate members. It creates and maintains bylaws and rules that govern the sale and purchase of deseeded cotton lint. These bylaws and rules are applied at big exchanges. The largest of these – ICE

Futures US – is located in New York and has been owned by Intercontinental Exchange since 2007. This is the exchange where the low- to middling-quality cotton known as Cotton No. 2 is traded. This cotton's standard bulk commodity characteristics are the global benchmark. A subscription-based information clearing-house business that specializes in cotton – Cotton Outlook – also aids the price discovery efforts and strategies of market participants. Finally, agribusiness and farming lobbies in individual states round out the big business institutions that create the cotton order. But there is more to the business of cotton than big business.

Alternative voices ordering cotton

The idea that cotton needs to be a large-scale affair in order to be "productive" and "efficient" certainly resonates around the globe. Many with big stakes in cotton continue to view this conventional approach to be the ideal. Planting vast swathes as single-crop monocultures with the assistance of high technology, and nurturing the crop with the standard package of petrochemicals, in this view delivers more of the white stuff for less work. Adherents to this perspective argue that technology-enabled, large-scale farming can reduce the amount of physical investment required to get the most out of each hectare. It is seen to minimize labor and other inputs per unit of output: the textbook definition of agricultural efficiency. And these methods have proven to be productive if standard measures of productivity, such as volume of output per hectare over single-crop years, are applied. For many, these facts on the ground cement the case for "going big" with cotton.

Other non-state actors with stakes in cotton have challenged this idea domestically and transnationally.[14] From the perspective of many small farmers, farmer advocates, civil society groups and smaller-scale businesses that seek better deals from cotton and ultimately to transition agriculture, the rationale for "going big" is fundamentally flawed. Individually and collectively, these voices have pushed back against industrial-scale production. Some, for instance, have posited that "bigger is better" is financially unsustainable. They point to the perpetual indebtedness of large-scale operations and also to the general financial reliance of massive farms on subsidies and other costly supports and relief that states have been willing to provide. In purely financial terms, then, the agribusiness ideal is not as cost effective for farming operations that embrace it as its advocates claim. On the other hand, in this light, agribusinesses that produce agricultural technologies and conventional inputs for sale are seen to be the primary financial beneficiaries. Big firms that produce tractor satellite GPS systems, modified seeds, or pesticides benefit when these items fly off the shelves. For firms that are publicly traded, consistent quarterly evidence of increased sales can help to push their share prices higher and, as such, boost the returns on investment their shareholders are able to realize. Securing farmer buy-in is consequently crucial for these firms and also for privately held firms that are not legally obligated to disclose as much financial information. Without it and the associated production of reams of public information on the "need" for scale, their business models would collapse like rickety derelict barns.

Alternative voices on cotton would tend to view the demise of "big cotton" to be a good thing. But their rationales are not simply financial. In the estimation of many advocates for land stewardship, cotton can be produced more sustainability by farmers and their families at smaller scales.[15] In making this case, they point to the growing body of academic and policy evidence that suggests that cotton systems that do not rely on conventional agrochemicals can and do succeed. Part of this success relates to the rapid expansion of the global market for cottons that have been certified by qualified independent authorities to have been farmed using more environmentally friendly practices. Approaches that fall into this category include organic certification systems and less formalized innovations rooted in "agroecology." The latter approach parallels organic in that it applies ecological principles to farming. But it relies less on the formal organic control and certification systems that some small farmers find to be too time consuming or expensive. Ideas associated with these schools of thought and practice endorse a radical descaling of reliance on inputs that originate "off-farm." For instance, organic systems encourage the use of materials that are available on-farm for fertilization, such as green manure or compost. Legumes that aid in nitrogen fixation, including green beans and cowpea, have also been introduced successfully in alternative cotton-farming systems.

Similarly, organic and agroecology approaches reject single-crop cotton monocultures on the grounds of sustainability. Instead, they embrace intercropping and encourage innovative crop rotations and fallow periods. These practices are undertaken to maximize biodiversity and total farm productivity over time. From this point of departure, it makes little sense to endeavor blindly to maximize yields of cotton each year. Put simply, the single-minded pursuit of that quest would ultimately yield lower overall total farm productivity. It would draw down soil fertility to such an extent that it would unsustainably undercut the potential of the land. As such, these innovators view the annual sprints that conventional producers make to get more cotton out of the ground than the year before as being fundamentally wrongheaded. The trail of exhausted lands that earlier runners have left behind in places as diverse as Texas and Tajikistan is considered by many to be indicative of the potential of conventional cotton to undercut soil health. And, in that area, organic and agroecology approaches are also associated with innovative practices. Many have embraced agroforestry ideas that aim to minimize erosion, such as the integration of trees with crops. In the case of cotton, such integration can reduce soil loss due to winds and uncontrolled runoff. It can also generate new income streams where planted trees bear fruit or nuts or can be tapped.

The application of ecological principles and organic methods to cotton is also associated with the development of viable pest control, fertilization, and rotation systems that can enhance the food security "status" of cotton-producing households. Chapter 3 engages directly with the controversial idea of a fiber/food trade-off and the limits of the conventional approach on food. While industry luminaries rail against this notion, beyond rotation crops, the reality is that whenever and wherever cotton is planted, food production potential declines. Cottonseeds can of course be rendered into oil suitable for animal and human consumption, and plants can be harvested, mashed, and formed into animal feeds. But if uninsured and unsupported conventional farmers in developing countries choose to rely primarily on cash earned from cotton to pay for their food needs, and they subsequently face production or marketing difficulties, hunger can loom.

Alternative visions for cotton offer farmers a more diverse range of food security "safety nets." Organic operations in Tanzania, for instance, have successfully intercropped cotton with a crop that traps pests and also yields a further stream of food and cash benefits. By peppering sunflowers through their cotton fields, they divert pests that would otherwise attack cotton bolls. And sunflowers that are used as trap crops – biological insurance – can subsequently be harvested. Even if they are slightly insect damaged, they will produce sunflower seeds and oil. These proteins and fats can be consumed by cotton farmers and their families, or sold to others for cash that can be used to buy food. But this is not the only pest reduction innovation that can yield enhanced food security. Farmers who apply lessons from the alternative school also grow neem or pyrethrum, or know where they can find these helpful plants. Both can be readily rendered into botanical pesticides. Cotton producers can consequently use these crops to develop new businesses. They can furnish themselves with safe pesticides and, if they have enough to sell, can help other cotton farmers to walk away from using dangerous petrochemical pesticides.

Taken together, alternative ideas about best practices for cotton that non-state actors advance matter in the world order.[16] Many individuals and institutions have articulated these ideas, including agricultural scientists working independently or with universities, global research organizations, or transnational consulting firms. Start-up cotton investors that have sought to buy and sell demonstrably different cottons have also been influential on a global scale. They have been a bridge between farmers, researchers, advocates, and consumers that want to see cotton done differently. Many new entrants in cotton have sought to extend knowledge of the alternatives to farmers, and to foster those alternatives through the creation of out-grower or contract-based farming schemes. To do so, they have collaborated with experts in agroecology and organic cotton, and have engaged in considerable "learning by doing." In India and in East Africa, corporate philanthropy enabled such experimentation and ultimately led to business success.

For their part, independent farmers have worked together to establish organic movements and best practice standards for new kinds of cotton. On the former, the International Federation of Organic Agriculture Movements (IFOAM) has played a major role in bringing supporters of organic cotton together across borders. National groups and movements have come together under this umbrella to discuss the development of their own national organic standards and certifications. And Textile Exchange, a non-profit formerly known as Organic Exchange and self-described "convenor, networker and catalyst," has helped to organize and inform organic cotton stakeholders with the aim of growing the global market. On that front, the alternative cotton market is now roughly thirty times bigger than it was when this non-profit commenced its operations in 2002. Before factoring in the economic value of the ecological benefits of organic production, in 2014 the rapidly growing market for certified organic cotton was valued at nearly US$16 billion.[17]

Non-state geopolitics and power over cotton

Taking stock of the non-state ideas and institutions that are now influential in the world cotton order is one thing. Accounting for the geopolitics associated with these ideas and institutions is quite another. Attempts to analyze the power that non-state actors exercise at the global level are inherently controversial. Some scholars in the field of international relations continue to be wedded to the notion that states are the only entities that hold power in the interstate system. However, when it comes to discussions of the transboundary politics associated with resources, political economists have long recognized the salience of non-state actors. And they have endeavored to show how and in what ways non-state actors can get other non-state actors to do what they otherwise would not do. Moreover, they have detailed many situations where non-state actors have and have not commanded the power necessary to influence the decisions states make at home and abroad.

Turning to a brief power analysis of non-state actors in the world cotton order, then, it is abundantly clear that big agribusiness firms have continued to exercise considerable direct power relative to other non-state actors and also to states. Business ideas about the technologies necessary to "do" cotton more "efficiently" and "productively" can be found in national farm and agricultural legislation and policy the world over. Well-endowed direct lobbying efforts and corporate public relations framing related to the supposed needs of cotton have consistently overwhelmed alternative voices. Nevertheless, the latter have in some cases countervailed the power of big business where and when they have been able to secure the acknowledgment of organic cotton in new statutes or policies.

But it is in the area of national and international standards and certifications development where the real non-state power politics has played out. In the absence of state-led efforts to develop rules and regulations pertaining to cotton industry best practices over the past decades, industry associations have stepped in to set the rules of the game. While national workplace health and safety and social standards and environmental regulations remain consequential, big businesses have invested in working together to set new standards specific to cotton production. In particular, they have aimed to develop new standards for on-farm production processes and methods. In so doing, they set out to regulate their own conduct and the conduct of the independent farmers that supply them in areas where government regulation had by and large not yet ventured. And they had a robust business case for spending the money and time to do so. In the aftermath of the textile and apparel factory "sweatshop" scandals of the 1990s, firms involved in the production and sale of raw cotton realized that they had a strong incentive to collaborate to ward off a similarly humiliating fate. They sought to build and present a friendlier face for cotton to consumers across the world and supplant the reality that cotton remained the principal source of global agricultural demand for petrochemical pesticides. Their efforts to avoid being painted with the same brush as those involved directly in clothing manufacture have now set new bars for more socially and environmentally friendly cotton production.

While this newfound corporate social and environmental responsibility might sound relatively innocuous, it is indicative of the structural power that big business has wielded over the cotton order. The major cotton suppliers, traders, retailers, and other stakeholders that financially supported and built this new CSR system labeled it the Better Cotton Initiative. Other non-state actors, including civil society groups such as the Worldwide Fund for Nature, facilitated this political project from the outset. And the power politics was evident when the initiative was first launched. After initial studies indicated that big businesses wanted to do "better" but to do so in ways that were less costly than organic certification, this initiative endorsed "principles" for the production of "better cotton," rather than hard rules. Principles in the global governance of resources have much less resonance than "rules" that are enforced by third-party agencies that are accredited to certify compliance. In essence, big business exercised its power to advance low-cost and light-touch principles that are not subject to legal or private-sector enforcement. The "Better Cotton" system they have built aims to structure government, public, and consumer comprehension of the practices necessary to produce cotton more "responsibly." As such, this system, detailed below in chapter 5, poses a direct challenge to the alternative cotton approaches that other non-state actors have endorsed.

That said, alternative voices have countervailed the structural power of big business to a certain extent. Public information related to the benefits of organic cotton has resonated amongst the growing global base of health-conscious and environmentally concerned consumers. Demand for cotton that is demonstrably "better" has pushed big-cotton retailers to also source cotton that has been certified by accredited third parties to have met or exceeded standards for organic cotton production. Many outdoor and high-end fashion retailers now offer numerous organic products and some are committed to solely sourcing organic cotton: Textile Exchange's so-called "100% Club." Alternative voices that support organic and other cotton systems subject to rigorous oversight, such as fair-trade certification, have chipped away at the direct, structural, and discursive power that big business has wielded. But do not just take my word for it. On Twitter, check out the work of advocacy NGOs Fashion Revolution and Trusted Clothes. These organizations, and other alternative non-profits and businesses, are collectively pushing back against the conventional cotton behemoths. They endeavor to reframe the language used to describe what is going on with cotton in the interest of genuinely improving the sustainability and equity of the global cotton order.

Forward into the cotton order

The rest of this book aims to bring the abstract contours of the order detailed above down to earth a bit. Any account of global resource or commodity politics would be incomplete without a lengthy higher-level overview of the type just presented. Thinking about the raw material that made up the shirt you are wearing, or the contents of your bathroom towel, is something some might consider to be overly self-regarding. But rest assured that it is much more productive than that. And it can even change your life and the lives of others when your thoughts turn to action, and you end up ordering organic pyjamas online. Or writing to someone or doing something directly about the horrendous wrongs that have been done in the name of cotton. The planetary scars of those misdeeds are patently evident in places such as the Aral Sea, a formerly large body of water in Central Asia that Stalinist planners infamously drained to irrigate industrial-scale cotton. To those who care to look, legacies of what has been done to people in the name of cotton remain evident in slavery museums across the Caribbean and Africa and in the United States and the United Kingdom. Industrialists in the latter country were the principal global beneficiaries of slave-subsidized cotton. Yet museums are not the only places where nasty human legacies are to be found.[18] These remain very real for women in the twenty-first century who must weed their cotton with hand hoes, or whose communities have been perpetually marginalized or stigmatized due to their association with cotton.

The crux of it is that the stuff that makes up the things we use every day is associated with a global politics and history. When it comes to cotton, this history and politics continues to be as deep or meaningful to daily life on our planet as inter-state warfare or international economic summits.

Over the past centuries, it has literally been under the pants of all politicians of world historical significance. And to restate for emphasis, for Europeans, control over pants – where they were made and what they were made of – was a key component of empire building in Africa and Asia. Europe's colonies kept cotton artificially cheap in the service of advancing industry at home. The application of scientific advances to this fiber also yielded some of the technological innovations that drove the Industrial Revolution. As such, cotton was literally one of the raw materials of the biggest story of dispossession in global history: the enclosure of the commons. Consequently, it was also central to the introduction of wage labor in the operations that the poet William Blake evocatively described as "dark satanic mills." And today this commodity is ground zero in geopolitical battles over how it should be produced, traded, transformed, marketed, and ultimately consumed, reused, or discarded.

Given the level of disorder associated with efforts to control cotton, it might be a bit of a stretch to refer to the framework within which today's geopolitics of cotton takes place as an "order." But the powerful states and non-state actors that have controlled cotton down to the present have done so in the interest of creating or bettering their positions in various orders: the imperial or colonial order, the advanced industrialized order, the Fortune 500 order, or the NGO pecking order, to name only a few. Applying the idea of world ordering yields an imperfect rough guide to the relevant contours of present-day geopolitics. And the analysis of ordering produced above is admittedly unwieldy and messy. But it is also necessary and meaningful to capture and re-present this imperfect snapshot. Cotton is about much more than a simple commodity market. Thinking about and mapping the big geopolitical picture or full spectrum of ways that people around the globe fail to connect, intersect, or run over each other trying to control a resource that we all use or see every day can be scary. But it can help us to understand the conditions that govern cotton and under which people come together, for better or worse, to bring about change.

CHAPTER THREE

Cotton on the Land

The political idea that market-oriented agriculture fundamentally rests upon the "freedom" or liberty that farmers have to make choices is globally consequential. Farming has been presented to countless numbers of introductory students as the textbook example of the benefits of free choices and free markets. From this point of view, farmers are assumed to be free to make decisions that they believe will advance their own self-interests. Many theories have trumpeted the "rational" efforts farmers make to choose and cultivate cash crops in the interest of maximizing their own utility or income. Praise has also been heaped on those that aim to bolster their profits each year through specializing and adapting to continually offer "in-demand" products. Powerful organizations, including the Bill and Melinda Gates Foundation, have portrayed agriculture to be fertile terrain for the development of capitalist enterprise. And when choices or markets fail, according to libertarians and other market fundamentalists, farmers and their families are always at liberty to pack it in, sell it off, and move to the cities.

Most of these political ideas are as unconnected to the realities of cash-crop agriculture as the American Dream portrayed in Hollywood products is to the daily lives of indebted consumers around the world. The flip side of rapid urbanization – de-agrarianization – is quite real.[1] But the rest of the prescriptive free-market ideal is just not happening. Agribusinesses and statute books consistently shape and constrain the choices that farmers make about what to grow and how to grow and sell it. Agricultural operations are also often far removed from the archetypal family enterprise. Industrial-scale factory farms and plantations typically make "choices" based upon government incentives, or shape the bounties and rewards that are on offer themselves. Big agribusiness companies also manufacture or generate demand not only for farm inputs, but also for new cash crops. And to top it off, growing numbers of farms have abandoned their quests for short-term profits. Instead, an emerging segment now literally puts the planet before profit, and farms the land differently in the interest of securing sustainable yields over longer time horizons.[2] Some have even pulled back from cash cropping itself and have embraced bartering and other approaches to exchange that do not rely solely on the money economy.

In the interest of elaborating a full spectrum of the politics of cotton on the land, it is nonetheless necessary to recognize the importance of farmer choice. Even though the choices farmers make are often subject to powerful control and sometimes to direct manipulation, choice is ultimately at the core of the making of this global resource. After all, unlike gold or coffee, but much like sugar, the potential natural substitutes for cotton, including flax, silk, wool, and hemp, are seriously competitive. The latter fiber is less water intensive and pest prone and can be successfully cultivated in nearly all of the present cotton-growing zones. As such, the choices farms and farmers continue to make to engage with cotton each and every year are where the geopolitics of cotton literally get going.

To draw out this politics, this chapter first offers a tale of two hypothetical cotton operations. This tale-of-two elaborates a situational analysis of some of the challenges that farmers working in very different contexts must navigate when they choose cotton, or are otherwise pushed, convinced, or coerced to do so. The first story commences through detailing the ideological rationales that underpin the consolidation of smaller cotton farms into factory-sized operations. It then focuses on the day-to-day decision-making context that average farmers confront in industrialized and rapidly industrializing farming economies where cotton is an option. Middle of the pack cotton operations in terms of size in those countries are, in a global context, relatively large operations. As such, the second contextual story focuses on the conditions that many operations with a much smaller economic footprint navigate across the Global South. And to be clear, despite their modest size, the direct and indirect economic impacts of the latter on immediate and extended families, communities, and even the earth itself can be vast.[3]

This comparison is then extended and applied to the specific topic of farm reinvestment decisions. Thereafter, the chapter continues to work with the tale-of-two heuristic to spell out the overarching fiber/food trade-off associated with cotton for both big and small producers alike. The concluding section then reviews the geopolitical questions pertaining to cotton on the land in light of the power, interests, and path dependencies that continue to intersect to reproduce this resource. Subsequent chapters spell out in greater detail the ways that big-picture politics associated with governments and global corporations infuse efforts to control and perpetuate cotton.

A tale of two farms

To be candid, recent popular and documentary films on the social and environmental implications of industrial-scale farming and the "industrial diet" can tell in a few images and graphics what a thousand scholarly words cannot. But do not dismay. There is something to be gained from revisiting this territory even if you have previously viewed *Fast Food Nation, Food Inc., Cowspiracy, A Place at the Table,* or other related films. If you have not, do check them out and push yourself to learn more about the movement that questions business-as-usual in big agriculture. And, alternatively, if you already are familiar with those critiques, think about them for a second. Because if you do you will assuredly remember that many films, books, and advocacy materials in this area have dedicated significant space to critiques of the place of sugar and its corn-based substitutes in the industrial food system. Perhaps unsurprisingly, in popular critiques, cotton has received much more limited attention than the power of sweetness. It has often only been raised specifically in the context of portraits of the ills of agricultural subsidies designed to socialize the risks, costs, and losses of relatively rich family farmers and factory farms.

But I digress. Tenure and business models for cotton in countries where industrial-scale farming is widely pursued are incredibly dynamic. Independent family farms that plant cotton using industrial technologies are less and less a part of the fiber-to-fabric story.[4] Farm consolidation is a reality that many remaining family-run operations either confront directly or have exceedingly proximate vantage points from which to make their own observations. And this wave of acquisitions has been especially consequential for cotton. The scale economies that can be reaped from cotton are relatively more enriching for big operators than those that can be squeezed from food grains. Cotton is a more expensive affair than grains when it is cultivated with the conventional package of agrochemicals and the latest tractor and satellite technologies. Putting aside the costly externalities that this approach yields for a moment, strictly from the perspective of big agribusiness, there is a devastatingly harsh case in support of consolidation.

If we suspend disbelief for a moment and put all of our eggs in the factory-farming basket, the case goes something like this. Family farms are simply too costly and too inefficient to survive. Why? Keeping and supporting a family on the land requires capital expenditures and investments that are not essential to the production of the commodity or good itself. Houses must be purchased, kept up, heated, and supplied with water and power. In this light, incomes earned by families that work with and live on cotton are not simply reinvested in the needs of the business. They are allocated to the real and perceived needs of the household, and also to the impulses and wants of household members: their "revealed preferences." From this obnoxious perspective, child care, university savings funds, the family car, and the internet subscription, not to mention Christmas and birthday gifts, vacations, the pets, and the kitchen renovation plans are all part and parcel of the small-scale efficiency problem. By living together on the land, family farmers appear to be responsible for sucking the life out of their businesses. They are not harmless toilers scratching out their existence from the land. In this view, they are profit-eaters that consume the life essence of their own enterprises, and of society as a whole when they perennially dine out on credit and subsidies.

Despite the reliance of bigger farms on even bigger handouts, this warped bigger-as-necessarily-better rationale for cotton does not end there. From a factory-friendly perspective, capital and technology-intensive approaches to cultivating this crop compound the supposedly inherent "inefficiencies" of family farming. Families that remain in this business take on relatively higher debt burdens than factory farms just to keep pace. Bulk purchases of the latest seeds that have been modified to be ready for proprietary pesticides that can also be purchased more cheaply in bulk are just the start of it. State-of-the-art mechanical pickers that harvest cotton bolls and immediately and automatically pack freshly picked seed cotton into market-ready bales are where farmer indebtedness really gets going.

When the new "module-making" pickers first appeared on the market in 2007, they were touted as a cost- and labor-saving breakthrough. In other words, through automating the packing of bales on one machine, they eliminated the need for farmers to utilize tractor teams to follow behind the pickers and bale up the picked cotton. Where the new machines were rolled out, old-style balers were rendered redundant, and so too were many migrant laborers who previously held seasonal harvesting jobs. Yet the carrot of potential savings was often not enough to convince small farmers to embrace the stick of further indebtedness. At the time of its launch, the new John Deere module-making cotton pickers were on offer for a cool US$600,000 suggested retail price. Competitive models produced by Case IH had a slightly less prohibitive price point, but were still seriously "big-ticket" items.

Advocates for increasing scale also offer other political rationales for their preferred vision for cotton.[5] Consolidation can enable bulk purchases of pickers and other equipment that facilitate fertilization, pesticide application, planting, and harvesting. And, if considerations of employment generation and sustainability are politically disregarded or simply ignored, the scale efficiencies to be had in the areas of irrigation, logistics, and storage can be seen to be considerable. On the former, small farmers in drought-prone cotton zones do not often command the market power necessary to reduce the setup, operating, and maintenance costs of sub-surface drip irrigation systems. While these systems can be more efficient and less costly than spray or surface irrigation, their prices and costs tend to require the extension of credit to family farmers. Similarly, scale can facilitate the logistics of spraying, harvesting, and marketing. Delays that family farmers tend to face when they rely on their neighbors for the equipment necessary to eradicate pests or get their crops off the ground can be reduced. Buyers, for their part, have enhanced incentives to head to the locations where the highest cotton volumes are available. Factory farms that organize storage for onward sale in centralized locations consequently reap serious benefits from scale.

Taken together, extreme arguments in favor of factory-style cotton farming tend not to be articulated at once as I have done here in the interest of emphasizing the underlying politics. They are generally offered up in piecemeal fashion, and in a much more subtle or less direct manner in business-oriented agricultural journals, quarterly reports, and board meetings. There also has been no fundamentalist grandstanding at shareholder meetings of the kind that occurred in Oliver Stone's film *Wall Street* when Gordon Gecko, the character played by Michael Douglas, immortalized the expression "greed is good." But even in the more nuanced forms they have taken, these views have been underpinned by a market-liberal assumption or prescription. Simply put, all agricultural labor is or should be a mobile factor of production. Families on the land are seen in this politicized light to be immobilized or inflexible rent-seeking labor. They use their resources – the very existence of their households – to obtain an economic gain from others, at the expense of others. Put another way, instead of being flexible price-taking labor, these small family businesses stand accused of holding back the potential "productivity," "efficiency," and wealth-creating potential of the business of big cotton.

Of course, the politics of all of this is highly contested. Terms such as productivity, efficiency, and for that matter "wealth" are subject to political contestation. Small farmers who have embraced different views about what each of these terms means and the practices necessary to advance them now offer the world fundamentally different models for cotton.[6] Their successes indicate that bigger is not necessarily better: there are equally sound rationales to abandon the new cotton factories in search of alternatives. After all, factory-scale farms are the biggest beneficiaries of costly credit and subsidy systems for cotton in places such as the United States. Even the most adamantly pro-scale cotton barons would be forced to admit that small farming families are a source of demand for agribusiness and for societies and economies more generally. Yet from a big agribusiness point of view, if these "small timers" disappeared from the land, the final frontier of the profitability of this resource would be within reach.

The good news is that all roads are not leading to a twenty-first-century rerun of the indentured or bonded migrant-labor surge that followed from the abolition of slavery. Suffice it to say for now that the idea of family farming in industrialized cotton-producing zones dies hard. Many small operators have undoubtedly benefited from subsidies that have helped to keep them afloat. And where and when those supports have not been enough, they have adopted new strategies to defend their livelihoods. Some have even walked away from their old business models in the interest of sustainability. That said, what does an average small farming operation for conventional cotton look like in the countries of the world where industrial-scale production remains the norm?

First and foremost, the debt pile on these operations is highly diversified. Some have inherited land and others have recently purchased more land or have leased it. Beyond mortgage burdens or lease rates, the typical small cotton farmer has access to a range of other credit products. Many hold significant credit lines with a mix of banks, equipment, and input dealers. These products are often extended over longer-term horizons and at higher rates of interest than the credit products that larger operators command. Beyond the enviable access they enjoy to materials and to finance to employ seasonal laborers, small operations tend to benefit from public or private crop-loss insurance. Government subsidies and payouts, when triggered by market or climate conditions, remain available to guarantee prices or minimize losses. Subsidies can also incentivize decisions about how much cotton should be planted, and where and how it should be planted. And they also ensure producers gain access to seasonal credit and to equipment when private credit provision is deficient, or credit scores fall short.

Farming is not a year-round endeavor. Farmers tend to have secondary businesses or salaried or waged employment in the off-seasons. And members of their immediate families are often not directly involved in cotton production unless they are of an age to drive tractors affixed with tills, planters, sprayers, or picking machines. Farm equipment and computers do most of the heavy lifting when it comes to the physical execution of production. The latter, often sold with or added to farm machines, offer applications that enable access to stored or real-time satellite images. These precision cotton technologies are now commonplace. They feature software that has been pre-programmed, or that can be programmed, to map specific soil, climate, and yield conditions. Numerous agricultural technology companies now sell smaller farmers complex yet user-friendly navigation and planning systems on credit. Sensors, monitors, and desktop geographic information systems provide their clients with images and calculations necessary to execute optimal seed placement, agrochemical usage, and harvesting schedules. And these technologies are accessible, even if some farmers continue to prefer the "old" manual in-field approaches to problem identification and yield maximization.

In terms of quality of life beyond their cotton monocultures, small cotton operators generally have access to water via wells that they treat through reverse osmosis or UV filters. If not, water is trucked in or potable water is purchased in bulk plastic bottles. Family farmers are either connected to viable electricity grids, or generate supplementary or all of their home and business power needs via off-grid solar panels or windmills. Residences are heated or cooled by systems that draw on these electric power sources, by geothermal systems, or by furnaces that burn piped or delivered petrochemical products. While roads to some farms remain unpaved, most enjoy access to maintained and paved roads within kilometers. Sanitation is effected through serviced septic tanks that have been certified to meet national and transnational environmental, health, and safety standards. And credit cards make possible the transactions necessary to procure the essentials of daily life such as food, clothes, and durable goods. Where and when these cards become maxed out, small farmers can access community and philanthropic support mechanisms. Food banks, thrift stores, and household reuse centers are close by, but many small farmers do not need to avail themselves of these services. They simply put purchases on a different card or credit line, or on their accounts or "tabs" with stores, coops, or other suppliers. Alternatively, they rely on their gardens or extended families and neighbors to make ends meet.

This cottony world of sugary soft drinks, satellite television, indoor plumbing, big trucks, and sports utility vehicles is nevertheless not an easy one. Managing small-scale cotton with big-time means is costly and time consuming. It also exposes farming families to considerable health and environmental risks. In countries that do not feature universal healthcare systems or low-cost health insurance, on-farm industrial incidents can be the source of serious health emergencies. The costs of treating arms that have been shattered by malfunctioning pickers can immediately undercut the viability of uninsured or underinsured small operations. Exposure to airborne herbicide and pesticide drift can yield the same result over the longer term. And in the context of climate variability and change, the "best" the intensive high-tech package has to offer might not be enough. When rains become more extensive or less frequent, and temperatures become more erratic, the risks and costs of this approach could become too much for farmers to bear. Societies that have afforded life-support mechanisms to small-time operations might be forced to consider cutting the cord. But until they do, the kids on the cotton farm will enjoy comfortable consumption, and the parents will have a relatively privileged, if imperfect, lifestyle.

But what of the day-to-day realities cotton farmers across South Asia, Africa, and other contexts in the Global South navigate where industrial agriculture is not the norm? To get at some of the politics of cotton on the land in those places, it is worth revisiting the US$600,000 module-making cotton picker. For that price, in December 2015, those with the means to do so could visit Alibaba.com – an online e-commerce portal – and order 150,000–180,000 long-handle hand hoes. At that time, it was also possible to order 70,000 mini-water purifier straws for about the same price. Or 11,000 very small solar-panel lighting systems, or 1,200 single-wheel, motor-equipped mini-tiller cultivators that could be used for seeding and fertilization. And perhaps even more surprisingly, including freight and insurance, at least 1,100 street motorcycles could be purchased directly from Chinese suppliers for around US$600,000.

At first glance, all of the goods above might appear to be incredibly useful additions to farm life in the remote rural cotton zones of places such as Burundi, India, or Niger. There is little doubt that these particular items in the specified quantities could make a difference in many lives connected to cotton.[7] If made available to the most marginalized and destitute farmers at cost through the extension of credit, or through direct giving, one module-making picker's worth of goods could clearly do a lot of good. But, on the other hand, it also could entrench underperforming systems and undercut the development of viable universal public services. Not all of these goods are cut from the same cloth when it comes to sustainable development. For instance, hand hoes tend to be a classic symbol of backward agrarian drudgery in places where many people continue to rely on hand hoes. Yet in the hands of weeding teams, they are serious sources of seasonal employment and community building.

Similarly, cheap water purifiers and solar-powered lighting rigs can and do prevent waterborne illness and help to bring people out of the darkness. But their mass provision in cotton zones could become a suboptimal substitute for more durable and universal public systems to sanitize and distribute potable water, and generate and deliver power. Mini-tillers, for their part, can be a big win for smallholders.

But if farmers have to pay exorbitant rental fees, or if communities lack small-engine know-how, the benefits are not as clear-cut. Like motorcycles, they rely on dirty petrol. Moreover, donkey power is significantly greener and more effective for smallholders with fiber to move at the small scale. These fantastic companions and beasts of burden can also command the "cool" factor of motorbikes where and when their owners take pride in their appearance and upkeep. And according to Alibaba.com, in the realm of international trade, live donkeys are literally priceless.

Life on typical smallholdings that farm cotton across the Global South may or may not be improved through the addition of any of the above goods. The difficult and unsatisfying answer is that there are no easy answers when it comes to addressing the development needs of small farmers, or their households and communities. But there is nonetheless an absolute certainty in this area. The portrait of a middle-of-the-road life with cotton in industrialized settings offered above could not be more remote from the daily lives of average cotton farmers in developing economies and the countries designated by the United Nations as "least developed."[8] And these disparities are an order of magnitude larger than the manifestly stark differences in farm size or debt piles. As regular farmers from around Bobo-Dioulasso, Burkina Faso have reported, after completing field visits to "normal" farms outside of Lubbock, Texas, everything seems bigger in Texas. Yet in terms of sustainability, bigger, more resource-intensive approaches do not necessarily radiate the right stuff.

So what does the other side of cotton cultivation look like? For starters, a regular farming family may or may not have formal title to the plot or plots that they farm. There is a strong probability that community gatekeepers and traditional authorities continue to exercise influence over land allocations and tenure rights. Assuming that our family secures land and tenure, and is able to live on or near the land to be farmed, each year they have to make two tough decisions. They have to choose whether or not to go with cotton, and, if the answer is "yes," what proportion of their land to allocate to this crop. Many seek to influence these decisions, including companies that buy or gin cotton, extended families, ministries, cotton-promotion agencies and locals with direct stakes in maintaining this business. Food and cash crops including maize, millet, sesame, sunflower, cowpea, and fonio are generally viable alternatives. So if our family chooses to go with cotton, they will hedge their bets and plant a diversity of food crops in addition to their cotton plot.

Water availability and access is an omnipresent concern in the generally remote and drought-prone zones where cotton is grown.[9] A water source such as a standpipe or stream is most likely to be located several kilometers from the home or plot, and cisterns are prohibitively expensive. And if the plot was recently cleared or is located on a hillside, the threat of flash runoff is all too real. Additionally, turning to household water usage, potability is not an ironclad guarantee. Members of the family will occasionally contract waterborne illnesses.

To gain access to cotton seeds and to the inputs conventionally used to grow cotton, our family likely entered into a contractual relationship with a cotton buyer.[10] So-called linked "input–output" contracts enable companies to provide inputs to farmers on credit in the interest of securing future cotton deliveries. Adverse weather events or the appearance of rogue traders or competitors at the farm gate tend to reduce the effectiveness of these instruments. And the reality that our family is not insured against crop losses is never good news for their buyer and can put the farm's income stream at risk. Our family remains largely on its own. The benefits of any meager subsidies that might be available tend to be appropriated by the largest and most connected farmers in their communities.[11] And the hope that they will receive timely advice or resources from the agriculture ministry or company to augment yields or address disease or pest infestations is typically slim to none.

In the fields, women and children are the primary weeders and crop protectors. To do so they use their hands. Plants are weeded with hand hoes, pesticides are sprayed by hand, fertilizers are hand-shoveled, mixed, and applied, and plants are watered via cans or old pesticide bottles. These cotton-related tasks are relatively more arduous than those that would be associated with other less input-intensive crops that the family could have planted. Cotton, in relative terms, is a crop that consumes more time. The associated time pressure tends to fall primarily on women. They bear the burden of daily food procurement and preparation. Water must be fetched, and wood or charcoal gathered or purchased. Chickens and the goats, oxen, or donkeys must be fed and watered and have their health needs looked after. Likewise, the needs of any sick or elderly members of the immediate family must be met. Beyond efforts to meet immediate human and animal needs, the house must be kept clean, and the thatched, tin, or corrugated roof and mud floor must be maintained. And latrines or other sanitation solutions need to be kept up or found.

Time-saving modern household conveniences are simply not a part of this story. Oftentimes, food and crop storage infrastructure is deficient or non-existent. Cell phones are however very much in this mix. Readily available refurbished or knock-off models can be a cost-effective silver lining for families that farm cotton to organize themselves to better allocate their labor and perform daily tasks. Yet even cell phones are far from a unidirectional positive. Enhanced connectivity pulls people off the land and toward the cities as reliably as it can bring people in farming communities closer together.

And much like most modern household conveniences, men tend also to be missing in action in this typical household scenario, especially in the African context. There is little doubt that most men "opt out" of many of the most physically demanding tasks associated with cotton, at least until harvest times when the prospect of cash sales looms. Where women do most of the work, men often aggressively seek to reap and control the rewards to be had at the market. It should go without saying that this reality raises serious household resource-allocation issues. Donkey hooves and household roofs can be neglected when earnings from cotton are reinvested in intoxicating brews and conspicuously consumed football jerseys.

Turning then to the reinvestment decisions that our two hypothetical farming families can face, the contrasts become even starker. On the debt and technology-heavy industrial operation, earnings from cotton tend not to be experienced as a one-off, windfall-variety type of cash injection. Most smaller farmers of this type have the financial means at their disposal to lock in the prices that they are paid, and all benefit from crop insurance. When the prices that farmers receive are credited as income and deposited into their accounts, business by and large continues as usual. In a non-crisis year, the amount earned was anticipated, and the family most likely borrowed against future earnings to smooth or enable consumption over the course of the year. As such, after the cash is in, it is time to pay down or retire old debt, and plan the tax, credit, and crop strategies for the following year.

This is also the time when reinvestment decisions are made. If targets were met and the outlook for the future is good, families can be tempted to reinvest more in the life and future of their households. Decisions about whether or not the university or retirement savings accounts should be topped up, or recreational amenities like in-ground swimming pools or new LED TVs should be purchased, are par for the course. When incomes fall short, or if they do so consistently, then it is possible that our family will feel pressure on the bottom line of their business. But, even at those times, the family will be able to maintain its standard of living. Failure still pays.

The biggest reinvestment questions that tend to crop up in this context do not always pertain solely to the physical capital of the farming enterprise but to the future of operations themselves. In areas where large corporate farms express interest in expansion, aging farmers with relatively smaller farms and depreciating capital assets can easily fall into the crosshairs. Staring down the prospect of costly outlays on new systems and the reality that they might not remain in the business long enough to make potential reinvestments pay, some farmers choose to walk away. Especially when immediate and extended family members express no interest in taking over the farming business. As such, some offer their lands up for larger farming enterprises to rent. Others take up offers from big enterprises and abandon their properties entirely. In so doing, they hope to receive debt-clearing payouts and build significant nest eggs. And they often do.

Turning to our other ideal middle-of-the-road farm in the developing world, there is little doubt that selling out into retirement is out of the question. While this family might indeed be forced to walk away and exit their lands at times of farming or financial distress, the bitter reality for most is that such decisions are not sweetened with an enduring cash infusion. Where formal title to lands is absent, or land transactions and values are subject to gatekeeping by powerful traditional authorities and formal officials, the prospects that departing families will receive a raw deal are all too real. And where more formal government-controlled land titling systems have been introduced, these have often created new problems for marginalized, disempowered farmers. In light of the potential family, kin, community, and government entanglements, many financially troubled farmers remain on the land even if some of their children do not. They fall back on the most enduring "safety net" associated with this crop. They simply walk away from cotton.[12]

While many success stories linked to cotton across the Global South have been trumpeted, the reinvestment reality faced by families that fall in the middle is grim at best. For those who have been strong, lucky, or connected enough to make this crop consistently pay, they have unquestionably reaped the much vaunted "white gold." But what does pay dirt look like? In a good year, a payout at the market in Togo equivalent to US$350 might be considered a middling financial success. And it must be emphasized that even this relatively meager sum can keep a rural family with a diversified farm afloat in that country. Yet the needs to be met are often immediate, and the consequences of poor choices tend to be quite dear. If the head of our household believes US$350 to be enough to keep things going as usual, for instance, and he also believes petrochemical pesticides to be at the root of this success, income would be allocated accordingly. Even if this idea bears no relation to actual crop protection needs and is unsustainable, especially in places where chemical company propaganda holds sway, the prospect that under-educated farmers will buy and utilize too many chemicals remains real. And this is scandalous, given the other pressing business reinvestment needs beyond the immediate basic needs of the household.

Potential reinvestments in the business can include implement and physical upgrades, such as the purchase of professional harvesting sacks that do not contaminate freshly picked seed cotton, and the construction of durable storage facilities that keep cotton dry. Resource pooling, in the form of community savings and credit groups, has in some cases enabled farmers in the middle to upgrade their modest equipment or facilities. But even these have not been a panacea in places where markets have started to "price in" periodic credit windfalls. Rural implement and infrastructure dealers in remote zones understand that they have a largely captive market. Moreover, for many in the middle, it can also make sense not to reinvest any extra cash in cotton-specific infrastructure. Hand hoes and tillers can be repurposed, and investments in food grain and maize storage upgrades can yield considerable household payoffs.

In the next section, we turn to the latter topic and also to the broader challenges cotton poses for food security. Our average family in the middle of the global bottom faces consistent pressure to "eat" its potential reinvestments. And the fiber/food trade-off does not stop or start there. It applies equally to industrial-scale conventional operations.

The fiber/food trade-off

Food security is an essentially contested political concept. There is no broadly shared transnational consensus on the investments or policies necessary to further its realization. That said, government officials signed off on an international definition of food security two decades ago at the conclusion of the World Food Summit. The Rome Declaration enshrined the idea that food security "exists when all people, at all times, have physical, social and economic access to sufficient, safe and nutritious food that meets their dietary needs and food preferences for an active and healthy life." As such, from 1996, the inter-state, international community agreed that food security was not solely about the physical availability of food or the lack thereof in famine zones. Food security, according to this definition, was also fundamentally about the capacity people have to access food day in, day out. Additionally, the realization of this status was linked to the access people have or do not have to sufficient quantities of healthy foods and to foods that they desire or deem to be culturally appropriate. The international consensus that food security has multiple dimensions now commands many supporters beyond governments, including activists, businesses, journalists, and research scientists.

Yet this "multidimensional" definition is by its very nature so broad or encompassing that it is possible for state and non-state actors alike to claim that almost any action taken on agriculture or food will necessarily enhance food security. Agribusiness firms and lobbies, farmers and farm advocates, and even professional researchers produce public information that links their operations, investment decisions, policies, findings, or alternative ideas directly to this concept. As such, there is a cacophony of contradictory information now available online and also in scholarly, peer-reviewed publications on the interventions necessary to realize food security. Businesses and governments in the Gulf region, for instance, have issued communications that link the establishment of new large-scale, export-oriented farms in the horn of Africa to food security "wins." Yet journalistic documentaries and professional reports issued by nongovernmental organizations on the impact of new land acquisitions in Ethiopia and elsewhere often make the opposite case. Similarly, while many academics decry the food security implications of these foreign direct investments, others continue to believe the erroneous idea that humans do not produce enough food to feed everybody on the planet.

Given these political realities, the seeds of confusion can easily be sown. And the contested politics of food security has considerable implications for cotton. Clearly, we do not eat this fiber, yet the world still produces enough food to feed itself. Voluminous research and an international consensus now back the claim that deficient distribution, grotesque food wastage, and insufficient social and economic access are at the core of hunger and dietary deprivation. So it might be tempting to conclude that cotton deserves a pass when it comes to food. However, its potential food security risks and costs are wide-ranging.

As land in more and more places has been devoted to cotton, more food has been produced while absolute hunger has been reduced. Of this there is little doubt. Yet the achievement of the food security aspirations encapsulated in United Nations sustainable development goal number two – *zero hunger* – hinge on the parallel need to feed a growing population more sustainably. And cotton, when farmed conventionally, is the crop that is the hungriest for petrochemical pesticides. It can effectively "eat" soils and "disappear" carbon sinks when old-school intensive methods are used, and where unsustainably extensive land clearance is ongoing. The food security implications are consequently crystal clear. To maintain all present cotton monocultures, arable land must be taken out of food production each and every year. Farmers make this choice willingly, or are convinced to do so by individuals and organizations with vested interests in the maintenance and expansion of cotton. In light of these interests and of the parallel rise of cotton cultivation alternatives, this annual fiber/food trade-off warrants further elaboration.

Organizations with direct stakes in cotton, including the United States crop promotion agency Cotton Incorporated, have produced public information on this topic that entirely discounts the notion of a trade-off between cotton and food. These accessible documents point to data indicating that less than 3 percent of agricultural land currently in use globally is devoted to cotton. They also emphasize the downward trend in the total global acreage under cotton cultivation evident over the past decades. And promotion materials are not wrong on both of these fronts. Cotton is now grown less extensively in industrialized cotton economies than it once was, even as it has come to be grown more extensively in places such as Brazil, Central Asia, and West Africa. But this information on the relatively small and declining proportion of arable land under cotton fails to underscore the petrochemical underpinnings of the intensification that facilitated this shrinkage in the first place.

Similarly, crop promoters rightly play up the fact that cotton is conventionally planted in rotations with food crops. But here too a key component of the food security story is missing in action. Many crops presently grown in rotation with cotton do not necessarily end up as food or in the nutritionless edible industrial products that food sociologist Tony Winson has labeled "pseudo foods."[13] Rather, rotation crops including corn and soybeans are increasingly rendered into ethanol and biodiesel. These fuels contribute to keeping the global economy hooked on the combustion engine. So, in a sense, instead of feeding the future, cotton rotations literally fuel an emissions status quo or at best incremental, path-dependent "change."

To be fully analytical, promoters are partly correct when they emphasize the point that new technologies can permit more fiber to be produced on less land in ways that are less harmful for farmers and that degrade soils less. The annual International Cotton Advisory Committee meetings in 2015 focused intently on the pesticide reduction imperative. And new drought- and pest-resistant seed varieties, and so-called "conservation" tilling techniques, have helped industrial operators to enhance yields and reduce erosion. Yet these technologies remain prohibitively expensive and are far from the only means people can employ to make cotton work better for sustainability around the globe. Labor-intensive cotton operations in developing countries employ and support the global majority of farmers, families, and communities connected to cotton. Many would crumble if they adopted the full package of industrial technologies and high-tech methods. Were they to do so, on-farm employment would shrink, debts would expand, out-migration would speed up, and corporate control over farming methods would be solidified. Moreover, food production would suffer. Out of perceived financial necessity, newly indebted industrial-scale operators could plant a less diverse mix of food crops than those they currently choose to cultivate. As such, dietary diversity could come under pressure as a range of nutritious and culturally important foods were dropped from crop rotations.

Promoters remain silent on these potential food security implications and on the ramifications of the conventional model and the consequences of monocultures more generally. The idea that conventional cotton does not take much land out of food cultivation and is congruent with sustainable development is a politicized house of cards. If applied in broad brushstrokes to developing countries, this perspective could prove the undoing of rural ways of life that keep people occupied. This eventuality is especially troubling, given the wave of unprecedented industrial automation currently washing over the global economy. As artificial intelligence renders a vast swathe of the global populace redundant and, as global food needs rise, a better balance must be found between fiber and food production. But crop promoters continue to equate "efficiency" with labor-saving approaches and "productivity" with short-term yield maximization. Across the Global South, there is no such thing as "disguised unemployment" or "ignorance" on smallholder farms that grow cotton. Rather, as discussed below, there is latent job-creation potential and unrealized educational opportunity. And both of these low-hanging fruit could spoil if vested interests disregard innovative alternative approaches to producing this fiber and to making it work better for food security.

To paint a more concrete picture of the food security implications of the fiber/food trade-off, it is worth returning to our hypothetical farms. A debt-fueled small industrial operation can absolutely contribute to enhancing the availability of food. If the farmer puts peanuts, corn, and cotton into a consistent rotation, there is a chance that the family could cultivate more than fiber and biofuels. Assuming buyers of food or oil-grade peanuts can be secured, then cotton, at least on the surface, appears to place less stress on food availability. Recall that cottonseeds can be rendered into oil, meal, and cakes that can be used to feed livestock. So, even when cotton goes in the ground, there is a chance that less land will need to be devoted elsewhere to producing animal feed. But cotton when grown conventionally in a rotation, as with all other crops, also needs fallow periods. And many operations of this type do not augment fallows through introducing additional periods under hay or pasture. The latter done in conjunction with free-range cattle can increase the availability of animal feed and of healthier grass-fed ruminants and fertility-enhancing green manures.

That said, the biggest looming challenge pertaining to conventional cotton and availability relates to the perceived qualities of regular rotation crops. High levels of agrochemical usage can undercut the perceptions buyers and consumers hold of the food-grade status of rotation crops produced on intensive cotton operations. As such, in the context of heightened attention to "no-spray" foods, corn and soybeans face an increasingly tough marketing context if they are not marketed as biofuels. With public disdain for petrochemicals continuing to grow, it will become increasingly difficult to characterize these crops as human-grade foods. Therefore, when it comes to augmenting the availability of animal grade and industrial pseudo-food products, conventional cotton is a sure bet. Rotations produce the raw materials for the corn and soy-based binders used in convenience and pet foods. But they make little contribution to increasing the availability of nutritious unprocessed foods.

Turning to accessibility and adequacy, the story is similarly bleak. Since industrial cotton farms produce a consistent stream of food crops, it might be tempting to believe that ever-bigger volumes of rotation crops necessarily yield lower food prices for consumers. But the reality of domestic food surpluses, subsidies, marketing controls, managed prices, and biofuel demand in many industrialized food systems makes it difficult to identify direct links between conventional cotton and enhanced accessibility to food society-wide.

However, cotton and rotation crop sales do contribute directly to the social and economic access that typical farming families have to food. And this is an area where further study of the food purchases farming families make could be incredibly fruitful. In industrial food systems, it is probable that our family relies increasingly on pre-packaged foods and less on their own-produced foods. Experts increasingly decry fatty frozen and boxed meals and sugar-laden confections and beverages on the grounds of health and nutrition. The chances are slim that farming families in industrial contexts that rely on conventional cotton do not purchase and consume significant quantities of these foods. They remain artificially cheap, despite their well-known knock-on effects. And the idea of imposing new consumption-reducing fat or soda taxes remains a political non-starter in industrialized cotton-producing zones.

Our other ideal cotton-farming family in the Global South can and does enhance the availability, accessibility, and adequacy of food proximate to its plot or smallholding.[14] Rotation crops can be sold to buyers that live or work in their community. When our family and others around it grow nuts, cowpea, or beans as rotation crops, mass sales can augment the availability of protein and lower market prices. Where bumper harvests are sold and get to markets where cheap protein would be otherwise unavailable, cotton rotations can have a positive impact on nutrition. Yet, to state the obvious, food security wins of this kind would be stronger if producers did not take fields out of food production to grow an inedible plant. The income our family earns from farm-gate cotton sales nonetheless generates demand when it is spent in local villages. As such, it can permit small traders in town to have more money on hand to access foods that they would not otherwise be able to purchase. Perhaps more expensive or nutritional foods. But here there are no guarantees. The brothel or gambling table could prove more appealing. This consumption effect also applies to the indebted industrial-farm family and has a similarly nebulous "it depends" impact on the realization of dietary adequacy.

If over-nutrition on industrial cotton farms is omnipresent, undernutrition is the constant companion of smallholder cotton. When our family chooses to plant too much cotton or faces blight, pest infestation, rain failure, or a host of other adversities, incomes will underperform and food purchases will suffer. Similarly, it is easy for this family to overshoot when making its annual commitment to cotton and to food and rotation crops. Needs are often immediate, and the prospect of a quick payout can convince farmers to reduce the size of their fallback food plots, especially when they are just getting by at the margins of subsistence. If an overly big commitment is made and food price inflation hits the market, the choice to rely on incomes from cotton for food can fail the food security test. At those times, the diversity of the family diet will be reduced and family members might even need to rely on the purchase of the cheapest available foods. In informal markets across the Global South, those foods tend to be the most degraded or hazardously unsanitary.

Taken together, these examples show that the fiber/food trade-off is real and much more complicated than crop promoters tend to suggest. More context-specific research on the food security situations of cotton farmers in industrial and smallholder contexts on this topic is required. A comprehensive program of genuinely disinterested research would enable us to map the totality of forms this trade-off takes around the globe. But there is little prospect that most organizations with vested interests in the status quo would be willing to fund time-consuming, costly, and delicate work linked to solving cotton's food conundrum. The perspectives of those that offer easy answers and simple solutions to this question tend to be reliably infused with heady politics. And an especially problematic politics is associated with the organizations that continue to deny that a trade-off exists. Much like our inability to rely on equipment suppliers such as John Deere or Case IH to play up any trade-offs linked to the technologies they sell, we cannot rely on the cotton industry to produce robust answers to its own problems. Independent research remains the order of the day.

On that front, to conclude with the silver lining, cross-national scientific research now suggests that alternatives to the conventional model can robustly supplant food security failures. Systems developed to support the uptake of organic cotton have challenged the fiber/food trade-off effectively in India, Africa and even the United States. Some of these benefits were recounted above in chapter 2.

A grounded geopolitics

Farmers choose to engage with cotton under global circumstances or conditions that they often do not control. Over the decades since the petrochemical-fueled agricultural Green Revolution, researchers, crop promoters, and marketing boards have expressed powerful ideas about how cotton could or should be "done." These top-down flows of know-how and advice have advanced the interests of the institutions or individuals that have articulated them. And farmers that were subject to this consistent stream of tutelage and were able to apply aspects of it also garnered the available rewards. Many who gained access to petrochemical input packages in this context were able to increase their yields and even better their lives.

Yet this conventional wisdom now faces strong headwinds. It can no longer be said with certainty that experts or members of the consuming public generally accept its core propositions. How farmers engage with alternative ideas, or with viewpoints on best practices that have been modified in light of the push back against intensification, is now serious transnational politics. But, as we have seen, farmers that sow these seeds continue to believe that there are payoffs to be had from excessively carbon-intensive approaches. These rewards might be fleeting, short-term, or contingent. But they remain enticing and largely under-regulated.

And it must be emphasized that producers and their families are not the parties most responsible for the persistence of discredited methods or for environmental despoliation. Farmers might create cotton, but their advisors, buyers, creditors, dealers, insurers, legislators, and regulators assuredly manufacture their decision-making contexts and operating parameters. The latter players cement a fiber-making status quo that would not otherwise exist. There is no preordained linkage between the fiber needs of the global economy and cotton. That connection must be actively maintained. In the case of cotton, it takes a good deal of politics to keep Pandora's Box wide open. Revealed consumer preferences for the high-end cottons that Egyptian farmers produce, for instance, are not set in stone. Were other fiber industries to command equivalent promotion resources or similar political wherewithal, consumers might instead favor other products – sheets made with bespoke Australian hemp or fair-trade Chinese silk, for example. Or linen shirts made with carbon-neutral Canadian flax, or organic wool socks from New Zealand.

Over the next decades, the politics of cotton on the land will only become more fraught. Farmers will feel pressure to dispense with their donkey teams and purchase motorcycle carts. And those that do will cede a key source of green manure and greener power. Similarly, equipment dealers will extend credit for the first time to many farmers in emerging and developing economies. The traditional rural life ways of those that take delivery of new tillers and tractors will be irrevocably altered as a result. Additionally, credit and subsidy systems that have kept families on the land in places where industrial farming is standard will be evermore strained and politically contested. And last but definitely not least, intensifying climate variability and change will challenge farmers in current cotton zones to think differently about food and its relation to what they choose to grow. But, as we will see below, gloom and resignation in the face of these looming adversities are non-starters. Hope can be found in the dynamism of the politics that now envelops and seeks to transform the world cotton order.

CHAPTER FOUR

Cotton for the Country

When states exercise control over cotton they project power over other states and territories. Historian Sven Beckert's magisterial book *Empire of Cotton: A Global History* offers the most comprehensive research published to date on this subject.[1] Coupled with the robust parallel work of social historians and international relations theorists on cotton, capitalism, imperialism, and industrialization, the historic record is now crystal clear. To glossingly reiterate a few of the now generally accepted key points alluded to above, imperialists and colonizers employed cotton to bring faraway lands to heel. Their efforts to keep cotton artificially cheap, including slavery and forced-labor production schemes in imperial domains and colonies, expedited innovation and industry in Europe. European states also controlled cotton and furthered their own national interests through excluding textiles or equipment produced elsewhere, or through incentivizing the homespun production of value-added cotton products. Given this history, cotton was front and center in the development of modern industrial capitalism *and* the structural economic asymmetries between territories that that system entrenched. As the era of mercantilist inter-state economic competition wound down, state-based competition to control cotton started spinning.

And it continues to spin. Today, countries that produce, export, or import considerable volumes of cotton still seek to use the state to control this fiber. Where and when they do, people connected to cotton elsewhere are necessarily disempowered. The United States, for example, has not become the leading export powerhouse through leaving the market alone. Rather, it has used its financial largesse and farm legislation to achieve enduring export-market dominance. Cotton exporters in West Africa and Latin America that would have otherwise been competitive, or that have been unable to fund or legislate similar supports, have suffered. Similarly, China employed state power over the past decades to build a massive national stockpile of cotton and realize its status as the world's leading cotton producer. Its thread and textile mills now draw down this "cheap" cotton while imports have been vastly curtailed. Control over cotton to advance the national interest has also been associated with exclusions beyond borders in this case. Taken together, government measures in support of cotton around the world were estimated to have reached an all-time high in 2015. According to the International Cotton Advisory Committee (ICAC), governments spent at least US$10.4 billion on subsidies to the sector that year.[2]

Yet the international spillovers or trade effects resulting from the exercise of government control over cotton remain only one part of the story of cotton for the country. As we have seen, the politics of cotton get going on the land. And it is there that states have found fertile terrain to actively and authoritatively manage cotton to advance their national interests. State-based priorities and directions continue to infuse many aspects of cotton production and marketing around the world.

This chapter seeks to capture and re-present the sheer extent of this persistent politics. To do so, it contends that in all places where cotton has been and continues to be produced, the "free market" has been entirely absent. Supply has been managed in ways that fundamentally depart from market liberal prescriptions. To grasp the depth and breadth of state control and the associated politics, the chapter commences through presenting the means that states have used to effect control. After an extended discussion on the specific ways that states create and perpetuate the cotton economy, the chapter then zeroes in on the rationales that states have employed to justify their activist approaches. The final section contrasts the present-day finance, trade, and direct investment challenges and opportunities states face when they rely on lint exports, and when they seek to add more value to cotton.

The extended focus on both the means and the rationales that states have utilized to advance cotton stands in stark contrast to the presentation of farmer perspectives and challenges made above in chapter 3. Here, instead of assuming and engaging with farmer choices and situations, readers are encouraged to engage with and assume the viewpoint of states. Trying to think like a state can be difficult. Many economics professors and contributors to the global business media encourage us each and every day to believe in the magic of the market. We are consistently exposed to information that implores us to think like capitalists, or, at the very least, like rational and acquisitive self-maximizing individuals. In this ideological context, those that focus on understanding and analyzing the roles that states play in economies each and every day tend to be inaccurately dismissed as protectionists. But even a cursory glance at the case of cotton indicates that the assumption that markets are self-organizing should be dismissed. Economic fairytales offer facile explanations and quick fixes. And, with cotton, none of the latter are available. There is no guarantee that a counterfactual reality – domestic cotton industries governed solely by forces of supply and demand – would deliver more benefits to societies or economies. And there is little doubt that big global merchants and agribusiness corporations with vested interests in cotton would in an alternate, more liberal reality do their utmost to prevent the emergence of a genuinely "free" market.

Thinking or "seeing" like a state does not mean that we necessarily need to agree with what particular states have done or continue to do in the name of cotton.[3] We should care about the "collateral damage" associated with production and marketing systems that have enriched well-connected farmers while impoverishing others. Morality matters. On moral grounds, it is imperative that the dispossessions and mass-scale despoliation that have followed from state actions are not forgotten and are redressed. But to truly understand the politics of cotton, it is appropriate, at least for the purposes of this chapter, to step back from big-picture twenty-first-century considerations of sustainability and human development. The practices and rationales that create and prop up cotton are rooted in older visions of the national interest and industrial development. And these visions remain relevant as states grapple with questions pertaining to the maintenance or emergence of employment-generating industries. In cotton-producing countries across Africa, Latin America, and South and Central Asia, population growth and rapid industrial and digital automation pose a considerable policy challenge. If government hands are already all over this sector, the prospect is very real that even more groping will be forthcoming as new labor-saving technologies become more readily available. Consequently, the good, the bad, and the ugly faces of state control must not be buried or actively imagined away.

Controlling minds, controlling the countryside

Much of the public information on cotton that is reported in the global media continues to be produced by agencies of states or by international organizations composed by states. Reports issued by the United States Department of Agriculture are closely watched, and so too are the ICAC's periodic communications on the market.[4] But this state-based intelligence is notable not only for the production, export, and price information that financial journalists tend to find headline worthy. This statistical information consistently reflects and entrenches the power states command over cotton. It does so through reproducing the idea that the most relevant knowledge about the cotton market should be disaggregated on a country-by-country basis. Most of the data that state-based organizations produce on volumes and trends continue to be presented in this fashion. And those that defend the statistical status quo rightly point out that there is a good reason for these organizations to stay the course. What goes on within states matters for other states and also for business.

Yet the road not taken with data is nonetheless more than a germane curiosity. Textile Exchange, the non-profit organic-cotton promoter, routinely produces public information on organic cotton buyers.[5] This information sheds light on a small slice of the global cotton market that would otherwise remain cloaked in darkness. If state-based agencies devoted an equal focus to disaggregating data on the basis of the corporations involved in production and trade, new light would be shed on the global cotton economy. And if data collection and dissemination also covered transactions of financial or risk-management instruments linked to cotton, more would be known about the contributions finance now makes to the size and growth of the industry.

As it stands, statistical business-as-usual does not offer much insight into the totality of cotton-linked transactions globally. Data are collected and presented in a manner that is convenient for governments and for members of the International Cotton Association that are subject to the rules states create. Corporate rule – the topic of the next chapter – is just not something that state-based organizations have been prepared to seriously identify, question, or chart for public consumption.

So states continue to influence many of the most basic ways that we think about developments linked to cotton. And this aspect of state control over cotton is only the proverbial tip of the iceberg. Financialization and trade and investment agreements have simply not arrested the development and use of the tools which states use to control this commodity and those that produce and exchange it. Some of the most common means that states employ to fluff cotton up nevertheless remain deeply controversial. States, acting individually and collectively, can and often do contest the legislation or policies that other states enact. As such, in the interest of making the most comprehensive and diplomatic presentation of state control possible, this section does not proceed to name or shame individual governments. Instead, it highlights a spectrum of ways in which different states have tried to make cotton pay and articulates some of the implications of those actions. At the outset, it must be noted that the interventions detailed below are not primarily rooted in Stalinism or state socialism. Far from it. Rather, the origins of heavy-handed approaches to cotton can be traced to the history of capitalism itself; to mercantilism, empire building, colonization, and postcolonial political economic development.

To commence with the obvious, the invisible hand does not deliver cottonseeds into the waiting hands of potential cotton cultivators. Seeing like a state, seeds that have the best chance of working have to be identified or developed, and they must also be delivered and take root. Few governments have left these crucial dimensions of the cotton economy to market forces. Instead, public funds have regularly been allocated to establish and build the capacity of national seed research and development centers. Public research has aimed to develop new cotton varieties that are better adapted to the soil and climatic conditions that prevail in specific cotton zones. These labors have in many cases borne fruit, enabling farmers to augment their yields and to also produce more consistent fiber qualities. The latter are an export-market essential. The qualities of many national origins would have been more erratic had governments not assured the uptake of the varieties that public researchers bred to be more dependable. Where the widespread use of more reliable public origin seeds set a higher standard, global buyers did not apply across-the-board export-price discounts. Free or subsidized distribution of new seeds to farmers also contributed to establishing and maintaining cotton as a viable income-generating activity. So too did the imposition of regulations to govern the varieties of cotton that could be planted, including where and how they could be put in the ground.

The public seed research, delivery, and regulatory story has not, however, been at the forefront of global interactions pertaining to the future of cottonseeds. Instead, proprietary seed technologies have animated popular attention, political activism, and public commentary. The uptake of cotton varieties that are the result of cross-species genetic engineering or new species-specific plant-breeding techniques nonetheless remains subject to thoroughgoing public control. National biosafety frameworks, committees and policies, ministries of agriculture, cotton marketing boards, and other high-level bodies and rules remain front and center. States still make the final decisions about whether or not field trials should be approved, or new biotech seeds rolled out. Corporations with an interest in promoting biotech seeds have typically sought to influence government decisions. And there is little doubt that the approval of transgenic cotton in particular places has been swayed by corporate propaganda, and also by the direct offer or prospect of licit or illicit rewards. But governments still command the power to govern seeds.

Several states have demonstrated their ongoing decision-making authority over the cottonseeds that are approved for use in their territories. Experimentation with transgenic seeds has, for example, been prominently reversed in Burkina Faso. Initial reports from the field in 2013 indicated that Bt cotton had permitted higher yields. But, by 2015 many smallholders there had become convinced that the new seeds were at the root of declining fiber quality.[6] While this deficiency is by no means inherent to genetically modified seeds, it was enough to convince Burkinabe authorities to endorse a return to the use of locally adapted and bred seeds. These had primarily been developed through conventional plant breeding, public funds, and development assistance. Governance or market failures unconnected to the seeds themselves might have been at the root of this decision – deficient training, for instance, or the lack of available or accessible fertilizers or untimely applications. Insiders could also have made this decision in the interest of securing another potentially enriching round of visits from well-endowed agribusiness lobbyists. Yet the fact remains that governments, not seed companies, make and control the investments and policy and regulatory decisions that get cotton in the ground.

Similarly, in many countries cotton boards continue to

exercise comprehensive power over production and marketing systems. These agencies govern the crop in ways that fundamentally depart from textbook free-market ideals.[7] At their strongest, cotton boards can control the allocation of exclusive seed, pesticide and fertilizer purchase, and import and distribution contracts. They also in some cases command monopolies on the provision of advice to farmers and impose rules on farmers directly or through agents. The latter can enforce bans on the intercropping of cotton with other crops, or meticulously apply stringent harvest and storage facility specifications. As market regulators, boards license buyers and regulate the activities of purchasers and sellers at the farm gate. Cotton boards also frequently specify indicative prices or price bands for seed cotton sales.

And they oversee the grading of seed cotton quality. Boards discipline unauthorized buyers and regulate the methods used to transport, process, and grade ginned cotton lint for domestic use and for the export market. Moreover, cotton boards can facilitate the international marketing of the national origin and lobby on behalf of lint exporters at home and abroad. The latter can be required when foreign buyers fail to honor their contractual obligations to pay specified prices in the wake of lower world prices. Additionally, the most influential boards help to facilitate the provision of credit or guarantees and are often tapped by industry to offer even-handed solutions to business disputes. They can even bail out failing firms or ensure their survival through "zoning": the practice of granting exclusive monopsony rights to individual buyers in specified geographic zones. Finally, boards produce information, organize seminars or public events linked to cotton, travel to the annual meetings of international organizations, and promote cotton production and use.

All of these past and current activities are deeply political. While they do not tend to be undertaken as a single "package," they persist in piecemeal fashion. In countries that lack cotton boards, ministries of agriculture, commercial regulators, and standards bodies impose aspects of this wide-ranging package. In so doing, they tacitly endorse supply management, a concept that was central to the development of all presently industrialized agriculture systems.[8] And today the political push back against the idea of domestic supply management has gained pace, especially in the negotiation of new regional trade agreements, such as the Trans-Pacific Partnership. The efforts of financially constrained newer countries to develop agriculture through the management of supply now hang by a thread.

Beyond this international politics, managerial approaches to cotton are associated with a profound local politics. Many systems have permitted city-based managers with elite political connections to effectively control what goes on in the cotton-producing countryside. Tender processes are a reliable source of enrichment for decision makers that succumb to the vice of corruption, and also for the politically favored businesses that are awarded fat contracts. In countries that command limited financial means, agents and inspectors that extend advice and enforce rules can be induced to "defect" from their responsibilities. If, for example, companies bribe rule enforcers at the market, farmers tend to pay the price. Scales that have been rigged to read inaccurate weights favorable to buyers can remain in use. Similarly, when money has illicitly changed hands, the physical identification of cotton quality might proceed in an overly exacting manner to reduce company payouts to farmers. Where companies buy regulators, cotton politics has exacerbated rural impoverishment.

But the role of the state in cotton does not end there. Transportation regulators, police, and customs and border authorities impose considerable discipline on buyers and prospective exporters. A truckload of seed cotton destined for a ginnery, or several containers on a lorry or train headed to the port, can be an enticing target for over-zealous officials that scent the possibility of a quick payout. And in countries where cotton is big business, the political aspirations or machinations of private cotton operators and cotton-board leaders tend to be on the radar at the highest levels of government, especially in states where clientelism or political patronage often undercut democracy. In authoritarian countries, when successful operators have directly supported opposition parties or indicated their intentions of running for them, their businesses have suffered. The recent case of Benin – a state that fundamentally depends on cotton for its hard currency earnings – is indicative. There, in 2013, a leading businessman who had made his fortune in cotton went into exile in France for over two years. He did so after facing allegations of being involved in an opposition-linked plot to poison President Thomas Boni Yayi.[9]

But we do not need to look to Africa to identify the enduring impact of big governments on cotton. The biggest producing and exporting states globally impose the most activist government measures in support of their industries. A 2015 ICAC report on the *Production and Trade Policies Affecting the Cotton Industry* provides the most up-to-date review of these interventions. State-centric data on this topic are enlightening. According to that report, the principal cotton-specific interventions are relatively standard. These include the offer of direct support to producers, border protections, crop insurance subsidies, and minimum price guarantees. China, for example, actively controls its imports in terms of volumes and values, and the China National Cotton Reserve Corporation manages the national stock mentioned above. It also applies quotas and sliding-scale duties on imported cotton. Growers there now receive subsidies for using high-quality seeds, and China also furnishes its farmers with direct support through paying them the difference between a target price and the average market price. According to ICAC, the United States, for its part, has converted many of its older, expensive supports to bring its cotton program into better compliance with the WTO statute book. In lieu of direct payments and countercyclical payments, the United States now provides its cotton farmers with a comprehensive crop insurance subsidy. When coupled with a new safety net program dedicated exclusively to cotton, known as the stacked income protection plan, these supports now help farmers to weather all possible difficulties except for their own negligence. In 2016, the US system continued to "distort" the world cotton trade while big cotton-farming firms prominently called for more handouts.[10]

Other countries afford their farmers a range of similar supports. Spain and Greece, for instance, offer their producers relatively more lavish subsidies than those presently disbursed in the United States. On a per pound basis in US cents, ICAC estimates that Spain and Greece provided average assistance equivalent to 44 and 39 cents per pound respectively in 2014/15. In the United States, owing to the revised policy mix and the absence of weather or other crisis events, the preliminary estimate of support was 6 cents per pound. Had a crisis hit, this number would have been significantly higher. Yet this level of support was itself relatively more lavish than the per pound subsidy farmers in India received that year. As a country, India boasts the largest number of cotton farmers and its national production volumes contend for world number-one status.

The Indian government now subsidizes its small farmers through means that are not specific to cotton, such as its comprehensive National Food Security Act. Farmers in Turkey and Colombia, for their part, likely took delivery of assistance levels that were equivalent to 27 and 24 cents per pound in 2014/15. Taken together at the world scale, production volumes were estimated to have held at around 26 million tonnes that year, while average levels of state-based assistance per pound shot up to 18 cents. If the global per pound figure holds true, it would be the highest level of assistance farmers have received since the dawn of the new millennium.

On the surface, then, it would appear that China's eye-popping estimated 58 cents per pound subsidy is at the root of evidently record levels of assistance worldwide. During a year when the world price mostly fluctuated between 60 and 73 US cents per pound, this costly system appears, at best, to be a little over the top. Yet the deeper reality is that China is simply the most direct about the ways that it supports cotton. Many governments now increasingly subsidize this sector through other means. The inter-state, international organization tasked with cotton is aware of this trend, but has not consistently produced estimates of the production effects or trade impact of these broader measures.[11] Historically cheap credit and universal or means-tested input and food subsidy policies or laws now create cotton as reliably as direct payments or at-the-border protections. So when it comes to cotton and the state, there is little doubt that much more is going on than allocations that target the crop specifically. Apparent or "effective" subsidies are a growth industry. Where they are available, universal sweeteners associated with redistribution and taxation make it easier for poorer farmers to take on the risks associated with cotton.

And it is hard to know exactly how much of a role these measures do play when farmers decide to go with cotton. Governments remain a crucial source of promotion materials, and campaigns. State-based cotton promotion agencies provide this business with the ultimate effective subsidy. They perpetuate the idea that cotton can and should meet the fiber needs of a growing global population. The influential US promoter Cotton Incorporated, for instance, is a state-run corporation. It is true that cotton farmers and importers pay for the operations of this promotion body. But they only do so because their outputs or imports are assessed on a per-bale basis. This assessment is levied by the United States Cotton Board, the body that oversees and administers the US cotton research and promotion program. The US Cotton Board administers Cotton Incorporated and in so doing is directly answerable to the United States Department of Agriculture. Crop promotion is consequently big government business.

Crafting national industries

Through seeking to control cotton in ways that they believed would advance their national interests, many states in the twentieth century attempted to turn this commodity into a raw material for national development. Where nation-building aspirations centered on facilitating the emergence of viable employment-generating industries, cotton was called upon. But, even today, as countries as diverse as Brazil, China, Uzbekistan, and Zimbabwe bend cotton to their wills, the lessons of history remain instructive. Cotton has in many cases been kept artificially cheap in the service of state-based efforts to craft industrialization.

In country after country, the cotton-producing countryside was consistently shortchanged before and after the first United Nations "Development Decade" was launched in 1961. Across Africa, for example, colonial systems that had been explicitly designed to maximize state revenues persisted as new nations emerged, joined the United Nations, and sought to transform or diversify their economies. Places such as Mozambique had previously been subjected to horrendous colonial systems of forced cotton cultivation.[12] And in newly independent Mozambique and elsewhere, the lesson that cheap cotton can be good for industry was not lost.

As the era of Asia's and Africa's formal political independence took hold, new governments in all places that had been subjected to cotton under colonialism did not have to look too far to learn this lesson. They simply had to go to the colonial offices that housed the officials that had been responsible for agriculture or cotton. In many cases, post-independence governments retained Europeans that had imposed the wills of colonial masters on the countryside. And once ministries and boards were fully staffed with nationals, colonial systems for cotton were not changed all that much. In some cases, they were subjected to efforts to assert even more centralized control. Massive farmer cooperatives that had previously been autonomous from colonial authorities, for instance, were sometimes turned into state-run or -controlled entities. State-run firms and parastatals subject to partial government control tended to manage all aspects of the business forward and backward from the farm gate.

Within and beyond Africa, governments demonstrated that they had an interest in keeping cotton cheap. In country after country, farmers were paid prices that authorities set to advance the national interest.[13] In places where states had picked up some or all of the tab for the procurement and delivery of chemical inputs, to reflect those expenses, payouts at the farm gate were set at lower levels. But payment levels were often much lower than they needed to be to make up for the range of costly public expenditures that kept cotton systems moving each year.

In countries where single-buyer environments prevailed or the state exercised total control over buying, ginning, lint sales, and exports, the political economic rationales for low producer prices were straightforward. On the one hand, if companies paid farmers the lowest possible prices in the absence of price competition, the costs of transforming cotton could be reduced. Ginneries – the places where seeds are removed from seed cotton and lint is baled – could realize better returns when farmers were paid less. Whether or not these ginning mills were in government hands or were simply highly regulated, lower seed cotton prices could effectively subsidize their operations. Even if they were abysmally inefficient.

As many governments during the postcolonial period also directly controlled or influenced the sale of ginned lint, they were able to allocate the downstream benefits. If, for example, domestic thread spinners needed cotton to be available at a lower price than the prevailing world price, governments that were willing to do so could ensure that this need was met. Lint subsidies or cheap credit from state-backed banks helped to make state-supported spinning operations more viable than they otherwise would have been.

Alternatively, and more regularly, governments marketed their national origin internationally in the interest of securing hard currency. Had farmers been paid fairer prices for their cotton – prices that were equivalent to a higher proportion of the world price – more of the export revenue stream of US dollars would have been directed toward the countryside. And, seeing like a state, hard currency was the sine qua non of industrial development for growing urban centers. Potentially restive populations in the cities demanded new opportunities for employment while governments continued to be subjected to the so-called "original sin" of the international payments system: the hard currency imperative. Foreign suppliers of the capital goods or equipment necessary for the transformation of cotton into thread, textiles, and garments were not willing to be paid in cedi, kwacha or rupee. They demanded dollars. As such, in the interest of self-preservation and development, many governments imposed a massive effective tax on the cotton-producing countryside. Cotton farmers were sacrificed at the altar of the almighty dollar for development.

Even after governments across Africa, Asia, and Latin America embraced more liberal ideas about rural prices and industrial development from the mid-1980s, the politics of controlling cotton to advance national interests persisted. Ideas about free markets were then ascendant. The International Monetary Fund and the World Bank brokered the supposed "solutions" of openness and austerity to heavily indebted developing countries. The Bank offered policy-based conditional loans, and many market-friendly structural "adjustment" credit packages were agreed. In this context, most single-buyer cotton systems were broken up, and pan-seasonal and pan-territorial approaches to pricing were generally abandoned.[14] Downstream, firms that had previously benefited from cheap cotton or other state supports were also exposed to import competition. Enterprises that produced garments with their own national origin fiber, and that had benefited from protection, typically ran into trouble or failed after supports were rolled back. But this was also an era notable for the massive growth of cotton production and exports. It was still possible for governments to make cotton pay for the nation. They simply had to control it differently to do so.

Rationales were revised in the 1980s in light of several massive changes in the international political economy. For starters, efforts to subject the world cotton price to active management had foundered. This price consequently reflected global supply and demand conditions for cotton, and not the aspirations of exporters that believed that an international system to stock and release cotton would better serve their interests. Given the concurrent emergence of strong competition from substitute fibers such as polyester, the failure of international supply management was doubly disadvantageous. Moreover, the simultaneous global interest-rate surge reduced the degree of financial freedom that indebted cotton exporters had to make the investments necessary to upgrade their cotton sectors. After the US Federal Reserve sought to tackle the inflation beast at home, the knock-on effect abroad punished many governments and firms in this business. Public and private debt and debt-servicing costs soared. And multilateral lenders asked troubled debtors to abandon some of the policy tools they had used to underpin their quests to industrialize.

As a result of these headwinds and also of the textile and garment export quotas developed economies imposed through the Multi-Fibre Arrangement (MFA), many cotton exporters had to step back from their industrial aspirations. Cotton for the country was downgraded. Everywhere in the developing world outside of the Soviet Union, from Brazil to Pakistan and in nearly all places in between, efforts to maximize foreign exchange for the sake of maximizing foreign exchange increasingly took hold. In this context, in country after country, strategies to rapidly scale up cotton-production volumes were embraced with varying degrees of success.

The go-it-alone, all-out pushes for foreign exchange that ensued ultimately yielded a global surge in the volume of cotton produced. This growth compounded the downward pressure that polyester had placed on the world cotton price. With the notable exceptions of China, India, and a few other producing countries where industrial policies persisted by hook or by crook, much of this new lint output hit the export market. From the perspective of states, to successfully compete in the global market conditions that dominated the 1980s and 1990s, the overarching imperative was to realize growth relative to what other countries were able to achieve. Governments knew that stagnant lint-export volumes, in the context of global volume growth and lower world prices, could only yield declining foreign exchange earnings. And the strategy they could pursue without fear of lender reprisal was seriously suboptimal. They could aim to beat the export market curve: the average rate of export volume growth that other lint exporters were able to realize. Growing sales volumes of the deseeded raw material to the world relative to other sellers was in many cases the sole available national route to securing an adequate share of the hard currency pie. And in hindsight, policy makers knew – or at the very least should have known – that this trade strategy was fundamentally immiserizing. In many places where it was embraced, it further entrenched weak terms of trade.

This sales orientation continues to fuel big-cotton politics in the present day. Farmers that were able to take delivery of enhanced packages produced higher volumes than they had previously from the same acreages. Many now expect more of the same and throw their support behind politicians and governments that they believe will best maintain the status quo. States, for their part, further entrenched the export volume path when they moved to introduce more business into this sector. As governments ceded direct control over the market and authorized volume-oriented private competition, they learned that the quest for foreign exchange did not always have to come at the cost of government revenues. In private hands, through the imposition of taxation, it could bolster state coffers and free up state resources. The latter could then be applied to correct market failures or to incentivize more production. This vision still animates many crop promoters across Africa and South Asia, despite the evident social and environmental limits of the conventional methods that have made volume-based approaches possible.

States remain hooked on the white stuff for context-specific and contingent reasons. Yet there are considerable commonalities. If cotton were to disappear from the countryside tomorrow in any of the leading producing and exporting countries, a short-term economic crisis would undoubtedly follow. Buyers that had entered into contractual relationships with farmers and with their own employees would be thrown into disarray. Any businesses that had been extended credit or credit guarantees, or that had signed off on forward sales, would be at risk of failing to perform on their loans or to deliver agreed volumes. Bankruptcies would loom and the multiplier effect would be curtailed. Employment that depended on the provision of services to consumers in cotton towns such as Surendranagar, India, would become unviable. And the industry and those that it kept going would inevitably push for bailouts or for the allocation of public funds to support those who had been directly retrenched or had suffered indirect job losses. Public and private insurance mechanisms would be triggered, and society would bear the costs. In a sense, then, the disappearance of cotton would seem to necessarily generate losses, reallocation, and costly public reinvestments as reliably as the closure of a gold or copper mine.

Yet, unlike those global resources, if cotton disappeared tomorrow, there would be many potential immediate substitutes. Bankers could furnish buyers with the right plans to encourage farmers to plant sesame and aim to feed the sushi eaters of the world. On their own, farmers could also plant millet for home consumption or for onward sale. Brewers that were attentive to local tastes could push farmers to switch to sorghum. And new businesses could emerge to encourage the cultivation of fonio, a nutritionally rich food that tends to be characterized as a "superfood," or hemp for the food and fiber markets. Cotton might in some cases be a big deal in particular places. But it is not the only potential game in a one-horse mining town.

The loss of cotton would endanger the interests that keep cotton going the most. Government jobs and short-term consultancies connected to the sector. Non-profit researchers, lobbyists, and promoters. Farmer organizations. Businesses that have taken on significant capital costs, such as ginneries, and other private operations directly connected to the fiber, including agrochemical pesticide and fertilizer purveyors. Any spinners that depended on local supply would need to procure cotton or substitute fibers from the global market. And those whose incomes, profits, and wealth-management strategies were linked to this crop would suffer. States have helped many individuals and institutions to make it with cotton and, as such, have built constituencies that faithfully strive to perpetuate this business. They spend massive sums on subsidies for farmers and continue to bear the costs of keeping this business going. Through their continual actions in support of this sector – their enduring commitments to cotton – states have deemed the potential risks and costs of altering the status quo to be too high.

Seeing like a state: cotton and industrialization

All countries that grow and export cotton seek to conserve the systems and economic returns associated with this crop. Cotton-producing and -exporting countries across the developing world have not attempted to steer fundamentally different courses. They have continued to play with the hands that they have been dealt, however meager those might continue to be. This conservatism stands in stark contrast to the recent risks that several developing countries that do not produce cotton have taken. The government of Bhutan, for example, has made a thoroughgoing commitment to transitioning all of its agricultural practices to advance principles of organic agriculture. No cotton-producing countries have made similar commitments. Their more conservative orientations necessarily prop up business-as-usual in the global cotton economy. Yet they navigate considerable changes and challenges in the service of re-creating this crop across the areas of finance, trade, and direct investment. It is consequently instructive to review the dynamic economic context that developing states confront when they attempt to export cotton, or to develop light manufacturing industries linked to cotton. Seeing like a state with direct interests in this crop or what can be done with it downstream can help us to better understand why states continue to put their faith in cotton.

In the area of *finance*, states now have many more degrees of freedom than they did a decade ago. There is simply much more public and private interest in funding initiatives or programs that can be linked to the development of agriculture. While the World Bank and regional development banks such as the African Development Bank continue to fund programs and projects that can facilitate cotton production and marketing, the relative importance of multilateral agencies as sources of development finance has declined. Many financial sector commentators and institutions now categorize developing countries where cotton is produced as emerging or frontier markets. Governments in places as diverse as Mexico, Ghana, and Turkmenistan can now more reliably tap the global bond market to meet their funding needs. They have been able to issue more sovereign debt at lower coupon or interest rates, and to use the proceeds from debt issuance to fund rural development initiatives.

A range of potential bilateral partners such as Brazil, China, and India have also been more willing to fund debt-based energy and infrastructure projects in developing countries. Numerous debt-heavy resources-for-infrastructure deals were struck during the global commodity "boom" of the 2000s. While credit remained cheap, in cases such as Cameroon, the roads that were constructed and power plants and ports that were improved or built were good immediate news for the cotton ginning, textile, and export business. The longer-term implications of increased government debt service and repayment commitments for cotton farmers and businesses were less straightforward. The new infrastructure might have come at the cost of future potential investments in remote rural roads, power grids, and delivery.

Moreover, since cotton now falls primarily into the hands of business, the new reality that stock exchanges from Lagos to Tashkent are emerging destinations for footloose portfolio investors has yielded numerous windfalls, at least over the short term. Where surges of hot foreign money have emerged and persisted, the size of the pool of investors that might potentially be willing to hold stock in cotton companies has grown. This in turn has reduced the risk that governments alone will be left to foot the bill in the event of private failure. Institutional investors such as banks and insurance companies, and endowment, pension, mutual and hedge funds, have piled into frontier exchanges such as Nairobi. Some have articulated an interest in remaining for the long haul. But for cotton-producing states with limited capacities to introduce timely and effective capital controls, the risk of a mass investor exit remains all too real.

The financial action has also not been limited solely to sovereign bonds or stocks, or to the frontier. From the United States to Zimbabwe, companies linked to cotton have also become buyout or investment targets for private equity firms and venture capitalists. And as the commodity boom has faded into the commodity bust and traders have fallen on leaner times, a new investor class has backstopped the sector. In particular, the state-owned investment funds known as sovereign wealth funds have continued to prioritize commodities investments.

The Singapore-based wealth fund Temasek Holdings, for instance, took control of Singapore-based Olam after this cotton-linked commodity trader was subjected to an attack by a short-seller in 2012. In the face of fizzling commodities prospects toward the end of 2015, Olam under Temasek was nevertheless able to arrange US$1 billion in debt-based financing to pursue an aggressive acquisitions strategy. Journalists writing for *Bloomberg* and for the *Financial Times* underscored the role of the state-owned fund in this case.[15] Similarly, as the global commodity trader Glencore simultaneously navigated troubled waters, the direct investment arm of Qatar's wealth fund expressed an interest in Glencore's cotton-linked agriculture business. In a sense, then, if we survey this scene like a cotton-producing state would, it is possible to see how the self-interested investment priorities of other states could dovetail with the national interests of those potentially on the receiving end of cash injections. But there are no guarantees that wealth funds offer first-best financial solutions for the states that are subject to their investment decisions. After all, they invest on behalf of the governments that control them.

Taken together, cotton-connected governments now navigate a more diverse range of options when it comes to financing the success of their cotton sectors. But the potentially fleeting profusion of finance on its own cannot assure export growth and industrial development. Both of those outcomes require states to bear the risks and costs associated with brokering ideas about new investment opportunities to potential investors. And the sources of finance that are most willing to allocate and disburse funds in the service of public priorities – the multilateral development banks – have had their own controversial priorities. The greater availability and diversity of finance has also introduced a range of new actors whose preferences might run contrary to the national interest, or to the interests of other financial players. Making cotton work for the nation in this financial context has necessarily become much more complex. Governments that seek to do cotton differently must secure a diverse range of financial-sector support. Yet there is no direct link between the views of those that express financial interests in the sector and the courses of action required for the sector to yield enduring or enhanced financial success.

Turning to *trade* and *investment*, real and potential foreign-exchange earnings continue to be constrained by the policies other states impose to support cotton. Approaching this issue as a state might, however, it is clear that the range of policy tools raw material traders have at their disposal to pursue industrial upgrading have shrunk. Multilateral rules and regional and bilateral trade agreements now inhibit recourse to the range of policies that were once associated with successful industrialization via import substitution. In the era of the WTO, regional trade and investment partnerships, and bilateral investment treaties, import bans and quotas are generally non-starters. But governments that seek to secure new industries can still do the heavy lifting needed to coordinate complementary investments and employ more limited policy measures to do so. Developing countries might not have the policy space that Taiwan or South Korea enjoyed as they successfully pursued the heavy-handed development of light and then heavy industry, but there are opportunities.[16] For countries that produce cotton, these have taken the form of more solid chances to add more value to cotton. The limits that the Multi-Fibre Arrangement imposed are now a distant memory. Instead, inducements or incentives for textile and garment businesses to set up shop in cotton-producing countries are now offered at multiple levels. Trade preference schemes, such as the United States African Growth and Opportunity Act, have incentivized the emergence of value-adding industries across southern Africa. In the interest of securing jobs, technology transfers, and hard currency, cotton-producing countries that are hungry for investment also offer a range of policy and contract-based incentives to potential direct investors.

But trade preferences, tax holidays, cheap electricity, infrastructure improvements, and the like have not enabled cotton to become a homespun affair in more than a handful of producing countries. China was the world's leading lint producer but was also the leading global importer before it started to draw down its national stock. The United States, as the leading global exporter, continued to export the vast majority of its deseeded cotton to feed Chinese mills and others across Asia. There, vast spinning and textile hubs have recently emerged in Bangladesh and Vietnam. While those two countries do not produce cotton, they have nonetheless rocketed to the top of the global cotton-import table. As such, cotton-producing countries are not the only countries able to claim location-specific advantages or offer effective investment sweeteners. Other leading producers, including India and Australia, export the bulk of their lint production. Only Pakistan and Turkey stand out as value-adding outliers in the top tier: they produce and also import significant lint volumes. Uzbekistan, Turkmenistan, Burkina Faso, Mali, and many other middling exporters have not achieved equivalent levels of inward or export-oriented upgrading. These states by and large remain suppliers of the industrial raw material.

The headwinds that most cotton-producing governments face on value addition do not stem solely from superior cost structures or economies of agglomeration in industrial centers such as Dhaka or Saigon. The latter are not set in stone. China, for instance, has seen the considerable out-migration of textile and garment firms as labor and regulatory costs there have continued to rise. The persistent obstacles also relate to the sheer volume of global garment production, to the thriving used-clothing business, and also to the global network of contract-based yarn, textile, garment, bedding, and towel suppliers. On the former, across Europe and North America, volume-based competition amongst discount-clothing retailers has been incessant. Low price points and fast fashion have enabled consumers in emerging and frontier markets more access to more types of clothing than ever before. Wardrobe, uniform, and inventory turnovers are now more frequent and, in this context, many cotton-producing countries have adopted more welcoming stances toward clothing imports. Made in China sheets, T-shirts, knock-off football jerseys, and jeans now fill open air markets from Abidjan to Dar es Salaam.

And the used-clothing business remains big business across Africa, Central America, and Eastern and Central Asia. In most countries where used clothing imports continue to be widely accepted, bales of used clothing destined for final consumption are now a more common sight at the ports than bales of cotton lint. The increased availability and accessibility of new and used garments in emerging and frontier markets has also fostered new consumer expectations. Many people now believe that clothing can and should continue to be low cost. Just ask your neighbor or phone a friend. Governments that seek to transform the cheap-clothing status quo in the interests of industrial development consequently face the potential encumbrance of a new, rising, and highly political social expectation.

Moreover, many cotton-producing countries confront significant barriers to entry into global supplier networks. Turnkey suppliers across Asia that were in on the ground floor of the sub-contracting and outsourcing boom of the 1990s are now themselves big brands. Hong Kong listed Li & Fung, for instance, is not simply a large-scale offshore sub-contractor. It is now a self-described global supply-chain manager and leading consumer goods design, development, sourcing, and logistics company. Along with its competitors, including the Spanish multinational Inditex, and the big branded clothing retailers, these firms now set the rules of the game for textiles, clothing, bedding, and towels. Cotton-producing countries beyond China can and do seek to increase their participation in these global supplier networks. But their participation on favorable terms is not assured. Branded retailers and suppliers tend to rigorously control product design and specification. As such, most new industrial-scale operations are subservient commodity suppliers. They are captive participants in the industries that use cotton and add the most value to it.

But, from the perspective of states that seek more from this industry, the future is not at all gloomy. The rise of value addition in Ethiopia, a marginal cotton producer, is instructive. Investors from China and across the Indian Ocean basin have taken note of Addis Ababa. And as governments across Africa work to develop a road map that aims to coordinate state-based efforts to do more for African cotton within Africa, this new hub, and others like it in Maseru and Antananarivo, will be closely watched. It remains an open question whether or not African governments will be able to coordinate industry players to serve local tastes. If they were able to do so, the production of cotton garments, sheets, or towels that Africans designed or branded could surge. And that eventuality would mark an historic departure from the model that has previously dominated that continent. The colorful cotton textiles known as African prints at present remain a primarily Dutch affair. The Vlisco Group has dominated this segment and continues to control design for its namesake brand from its headquarters in Helmond, the Netherlands. Yet, even as this firm has sought to rein in knock-offs and copyright infringement, it has ceded control over design at its subsidiaries to Ghana and Cote d'Ivoire. Little would stand in the way of governments that were willing to collaborate to turn this one-off into a trend.

In sum, then, the global cotton economy is still fundamentally about what states do or think they need to do to make it up. There is no necessary connection between cotton and the actions that states take to earn hard currency and employ more people. Other natural or man-made fibers could deliver industrialization and economic transformation, an aim that states share and widely articulate. But states that rely or depend on cotton have a mutual interest in keeping the global economy threaded to this needle. Evidence that alternative approaches and artisanal-scale cotton-textile industries might under specific conditions more reliably deliver sustainable development has as yet failed to sway the commanding heights.

CHAPTER FIVE

Cotton in Company Hands

If states continue to animate the front-page, big-picture international politics of cotton, corporations bring the politics of cotton to life each and every day. Companies connected to cotton clearly do not enforce pan-territorial agriculture and trade policies. They obviously do not set standards for quality related to the physical grading of the fiber. And they certainly cannot facilitate inter-state initiatives, discussions, or the resolution of inter-state international disputes. But they do benefit from such international engagements and inform the associated politicking. Their preferences and interests are key considerations in all government actions taken in support of cotton worldwide. Moreover, corporations have been central players in the political development of a raft of new voluntary principles and rules that aim to govern production methods. When researchers and advocates in civil society have blown whistles on poor practices, and consumers have demanded change, firms have been the primary political targets. And, in the wake of all of this new political action and attention, new kinds of companies have emerged and taken flight. Today, corporations that offer advice on social and sustainability performance, or that assess or craft public relations materials on these matters, make key political contributions to securing the future of the global cotton business for business. So too do financial firms that specialize in commodity "risk management."

The latter term is fashionable industry jargon for narrow financial planning that addresses the perceived immediate needs of firms, and not the full spectrum of political economic risks associated with cotton in company hands.

In the twenty-first century, the involvement of privately held or publicly listed profit-oriented companies with cotton is not inherently legitimate or illegitimate. Far from it. And corporations that buy or sell cotton know this to be true. Even if they are unwilling to admit as much directly, or if they fully adhere to the letter of the law in the jurisdictions where they operate. A cursory review of the web-based communications of the leading firms can reveal their newfound attentiveness to the underpinnings of the cotton business. But do not take my word for it. Just drop this book and do a simple internet search of each of the leading global cotton traders: Louis Dreyfus, Cargill, Noble Agri, and Olam. Simply type those names into your favorite search engine and include the additional search terms "cotton" and "sustainability." From those searches, you would quickly surmise that the politics of generating demand for cotton and pushing back against market-leading polyester now fundamentally rests on corporate communications that trumpet supposed sustainability qualifications. Firms that seek to control the future for this fiber understand that credibility is their new currency. And, to stick with the analogy, this currency is not centrally issued or regulated. As such, it remains subject to widespread counterfeiting, and the potential for debasement is all too real. State-based regulators have ceded this territory to industry self-regulation at the same time as multiple approaches to realizing "sustainability" have thrived.

Cotton companies make wide-ranging contributions to the governance of this commodity. And, when businesses intervene in the realm of governance, their actions can enhance and also detract from the perceptions that other stakeholders in cotton hold of their activities. To understand the new political terrain where corporations seek to secure legitimacy, the framework advanced by the political theorist Jan Aart Scholte is instructive.[1] Scholte has analyzed the efforts that non-state entities such as civil society groups and corporations make to shape transnational or global governance. He has done so in light of the reality that companies and NGOs are not inherently legitimate contributors to the development and operations of problem-oriented global governance institutions. As they do not command territories or militaries, and they do not establish the rule of law, non-state political actors are precarious participants in transnational governance. Multi-stakeholder global governance is itself not inherently legitimate or illegitimate. Initiatives such as the United Nations Global Compact to facilitate "better" business practices remain voluntary. And inter-state, international organizations can at their discretion choose to reject participation or collaboration with companies or non-profits.

In this light, the continued engagement of corporations in governance innovation at the global level rests upon business efforts to secure legitimacy in the eyes of states and societies. If companies do so, they can simultaneously contribute to enhancing the legitimacy of the governance institutions or initiatives that they seek to advance. Applying Scholte's framework, then, there are several ways that corporations can bolster perceptions of their own legitimacy and also the legitimacy of the governance institutions they participate with or work to build. The fact that there are multiple routes to securing legitimacy stems from the reality that this concept has multiple dimensions. These dimensions are not mutually exclusive: perceived failings on any given dimension of legitimacy can significantly undercut firms that seek to maintain or enhance their social buy-in or "license" to operate.

First and foremost amongst these dimensions is legal legitimacy. Corporations can secure legal legitimacy when they play by the rules that states establish or are at least perceived by states to do so. When they overshoot or underperform as a direct result of gaming the system, or fall foul of statute books or state-based regulations, they do not command legal legitimacy. Mismanagement or fraud that results in fines, jail time, insolvencies, or bailouts, is by definition the antithesis of legally legitimate corporate conduct. On the other hand, political actors can be seen to secure *democratic legitimacy* when they demonstrate that they are accountable for their actions. If companies actively redress public perceptions of corporate wrongdoing and work to prevent the recurrence of the practices that were painted as illegitimate, they can present themselves to the public as good corporate citizens. While this accountability dimension of legitimacy is often applied to those in public office in democracies, transnational corporations involved with cotton increasingly recognize that stakeholder accountability is a business essential. Firms that operate in authoritarian countries or in established democracies know that members of the global public can now instantly hold them to account via social media. Without notice, their business operations, political lobbying, and industry-building efforts can be lauded or exposed.

On the other hand, the effectiveness dimension of legitimacy underpins the place of private enterprise in the global cotton economy. Businesses need to exude *performance legitimacy* in order to continue to operate. They must foster the perception that they can deliver on what they have set out to do. So if they are publicly traded, and aim to maximize shareholder value through beating the average rate of return industry-wide, they need to be seen to be striving to do so. Similarly, if they desire to change the rules of the game for their business or for the industry as a whole to better align cotton with sustainability principles, effectiveness is the overarching consideration. In the latter case, public recognition of failure to secure performance legitimacy would undermine much more than the bottom lines of the corporations concerned. That eventuality might pull the rug out from under broader business efforts to entrench voluntary rules that might have a lighter touch or come at a lower cost to firms than genuinely public state-backed laws and regulations.

A further and increasingly important dimension of legitimacy is morality. Corporations that support the business case for sustainability leadership or that face the court of public opinion need to be perceived to be moral actors. Business contributions to non-state or multi-stakeholder governance initiatives are simply more credible when they appear to indicate that firms are pursuing transcendent values. Corporate operations can also yield *moral legitimacy* when objectives that the public views as noble are rigorously pursued. This new frontier – corporations as moral actors – has not been universally embraced by businesses connected to cotton or by members of the global public. Some still claim that morality in business is suspect, undesirable, or even impossible. But many businesses directly or through foundations seek to be a social conscience. And of late when corporate morality has prominently and publicly failed, as occurred with the collapse of the Rana Plaza garment facility in Bangladesh in 2013, business closures have promptly ensued.

A final essential dimension of the legitimacy of business pertains to social inclusion. Any business that actively fosters social exclusion through paying employees or suppliers, directly or indirectly, wages or fees that do not enable family life now risk being publicly shamed. In cotton, this dimension of business legitimacy is closely linked to morality. Yet it remains distinct from it insofar as corporate contributions to social inclusion do not necessarily relate solely to the moral conduct of companies. Social inclusion is about the capacity of firms to foster dignity and the fuller, more equal participation of stakeholders in society. And as with all of the other dimensions of legitimacy, how social inclusion is assessed and portrayed by companies or by their critics is ultimately political.

But no serious analysis of business and cotton can be devoid of politics. Businesses get political when they aim to foster the social perception that their activities and interests are indeed legitimate. Analyses of companies in cotton that assume away the politics of doing business do so at the risk of obfuscating the most dynamic area of development and change in this industry. Cotton companies of all stripes are now explicitly political actors. They advance a diverse array of new norms. And they do not do so equally or even in the same transnational alliances. The reality of political competition to establish new best-practice standards is nonetheless only one part of the politics story. The daily operations of firms in this business also tend to sanction or legitimize political economic ideas or practices that are by no means permanent features of the global economy. This has particularly been the case in the area of finance. The gamut of financial risk-management tools currently in use are by no means optimal or enduring. They have failed at a considerable social cost before. And the next global financial crisis could afford a ready opportunity for public authorities to reform finance in the interest of ending the private appropriation of the available benefits.

This chapter proceeds to analyze the global business of cotton through the lens of legitimacy.[2] To do so, it focuses first on the financial side of the global business, and then on physical or "real" business activities and initiatives in support of cotton. At first glance, this separation might appear to be artificial or polemical, but it is assuredly analytical. Physical cotton undoubtedly requires finance to get going. But it is by no means clear that it needs financialization. The global cotton economy reproduced itself and adapted for centuries without "innovative" finance. Each section below focuses first on democratic legitimacy and then sequentially on performance legitimacy, moral legitimacy, and social inclusion. The concluding section reflects on the politics and power of business in the geopolitics of cotton.

Financialized cotton

It is one thing for global cotton traders to raise money in corporate debt or equity markets, or to take delivery of cash injections from private equity investors or sovereign wealth funds. It is quite another for these firms to engage in risk-management strategies that aim to make their prices more predictable and incomes more reliable.[3] The latter can sound rather innocuous and has been portrayed by business as a new fundamental. And it is true that hedging to lower the risk of nasty price movements can keep liquidity – the wheels of cotton commerce – flowing.[4] But at present the scale and scope of the new financial "planning" has amplified volatility while fluffing up the potential earnings that can be squeezed from this fiber.

This outcome is not inherent to instruments such as futures contracts, nor related solely to the positions that traders with physical cotton move to take in the market to cover the risk of loss. The absence of market rules, such as position limits and the development of automated high-frequency trading, has enabled financial speculation on a mass scale. It is now possible for Cargill, for instance, to use the new financial tools to do much more than hedge its positions in the physical cotton market. Through one of its numerous financial subsidiaries, it can employ split-second trading strategies that place bets for or against futures prices in the interest of maximizing short-term profits. Such speculative activity is about as unnecessary for the delivery of physical cotton as snowblowers are to the daily lives of the inhabitants of Cotonou. Budgeted operating expenditures and short-term trade finance loans are the standard means used to deliver the real goods. As such, the world could easily get on without the fluff if public authorities reined in firms that currently engage in speculative proprietary trading. And if they closed the door on market players and margin traders that speculate on cotton without holding any physical fiber, things would be a lot more real.

Yet the current politics of financing the cotton trade do not start or stop with financial innovation and its impact on market participants or societies more generally. Louis Dreyfus and Cargill, as privately held firms, are not subject to the same financial disclosure requirements that publicly traded companies face. The chair of Louis Dreyfus Holdings, the billionaire Margarita Louis-Dreyfus, is a particularly strong advocate for family ownership. In her estimation, private ownership offers superior flexibility to businesses insofar as it does not subject firms to the "mood of the market" or redirect their activities to the "short-term delivery of profits" to shareholders.[5] Given this reality, less is known about the prices and costs associated with the real and financial commodities businesses of Dreyfus and Cargill than is known about companies that actually issue stock to the public. And their vast global networks of offices and subsidiaries are as a consequence highly opaque. The prominence of the latter status in many commodities businesses has of late attracted considerable policy attention.[6]

Where financial transparency is lacking, for example, subsidiaries of the same private parent firm that are registered in different countries can engage in so-called "transfer pricing." Under those circumstances, any prices that subsidiaries agree to sell goods or services to one another at do not need to be rooted in broader market conditions or prevailing prices. They can reflect corporate priorities, such as paying less tax in higher-tax jurisdictions. Subsidiaries registered in such places can engage in intra-firm transactions with subsidiaries registered in lower-tax locations that aim to generate paper losses and reduce their tax burdens. And this questionable practice is just the dull edge of the knife of non-transparency. Beyond cotton, commodity supply networks have increasingly been called out for trade mispricing and mis-invoicing, and also for enabling so-called illicit financial flows: the clandestine funneling of questionable cash out of commodity-producing countries.

And it is not necessarily the case that the publicly traded cotton giants are "cleaner" than the privately held cotton titans on all of the above. Noble Group is headquartered in Hong Kong and registered in Bermuda, while Olam International is registered in Singapore. All three locations are noted secrecy jurisdictions or tax havens. Moreover, Noble's agricultural business, Noble Agri, was until late 2015 run in partnership with the China National Cereals, Oils and Foodstuffs Corporation (COFCO). In December of that year, Noble announced its intention to sell its remaining stake in Noble Agri to a COFCO subsidiary. To put it lightly, Chinese state-run firms like COFCO are not at present notable for their voluntary openness to more transparent business practices. Coupled with the reality that Olam's new majority shareholder, Temasek, is a sovereign wealth fund controlled by the government of Singapore, there is little to suggest that greater transparency will soon become the new normal for cotton. But a future crisis could certainly undercut the legitimacy of these businesses and devastate the suppliers that they and other much smaller cotton merchants rely upon for the stuff that keeps the mills moving.

So thinking about *democratic legitimacy*, then, the starting points are less than auspicious. Cargill and Louis Dreyfus have been at the forefront of the campaigns which commodity traders have waged to derail the imposition of any new rules that would establish limits on the positions that they can take in futures markets.[7] These firms claimed in 2014, for instance, that their global buying and delivering businesses would suffer if the US Commodity Futures Trading Commission imposed position limits. The proposed rule would have forced firms to certify that big short or long positions that they took in cotton or in other futures accurately reflected their risk-management needs. In particular, they would have had to verify that any seriously large positions were in fact hedges against the real risks posed by their physical inventories. And this was a level of oversight that the industry could not stomach. If realized, it would have forced major traders to revise business models that had come to rely on the periodic financial injections that can result from the successful execution of short-term trading strategies. And to reiterate, the major traders argued that position limits would undercut the real business of commodities, a business, it must be noted, that they had profitably engaged in for decades before their profits became hooked on the steroids of speculation.

Similarly, when Switzerland considered imposing new transparency rules on the industry in 2013, cotton traders there also did their best to maintain business-as-usual. Many commodities contracts and deliveries arranged or transacted in Geneva, Zug, or Zurich do not aim to bring the raw materials concerned to Swiss soil.[8] Yet Switzerland has come to play a central role in the global commodity trade. Its status as a commodities hub has been underpinned by a favorable tax regime and dedicated infrastructure, including specialist bankers, consultants, and law firms.[9] So when a membership-based Swiss NGO known as the Berne Declaration pushed the Swiss government to impose mandatory transparency on the industry in the interest of curtailing illicit practices, traders stood their ground. In particular, the Swiss Trading and Shipping Association publicly stated that it was willing only to discuss the uptake of voluntary measures. And the association's argument that Switzerland needed to protect its role in commodity trading ultimately held sway, despite the fact that there is no geographic rationale for Swiss prominence in this business. As financial transparency expert Alex Cobham of the Washington, DC-based Center for Global Development has noted at length, opacity has been Switzerland's primary comparative advantage. Industry luminaries greeted calls for greater transparency there with the ominously veiled threat that this business might find more fertile regulatory terrain in Singapore. As such, in this case and in the case of the United States, cotton traders implicitly considered democracy to be a dangerous risk. This stance bears directly upon their credentials in the area of democratic legitimacy.

Turning to effectiveness or *performance legitimacy*, cotton merchants have recently worked together to make the futures market more effective for traders. In 2014, market participants with stakes in cotton futures agreed on the parameters for a new global benchmark futures contract.[10] As rigorously reported by Greg Meyer, the in-house cotton expert at the *Financial Times*, the world's leading business newspaper, previous US rules were seen to pose unnecessary impediments to the physical trade. In particular, under the old rules, any cotton tendered to fulfill a US-listed cotton futures contract had to be sampled and graded by the United States Department of Agriculture. This rule created headaches for traders who wanted to hedge risks in the world market beyond the United States. At times, they reportedly had to scramble to find US-origin bales to meet their contractual delivery commitments. Under the new rules agreed for the global benchmark cotton contract to be listed on the Intercontinental Exchange (ICE), cotton from numerous countries including Australia, Brazil, India, and elsewhere can now be tendered. Deliveries to several foreign ports including Malaysia and Taiwan are now possible, and laboratories in other countries are authorized to grade cotton.

There is little doubt that all of these reforms could make the physical trade more effective. They will assist firms that need to locate and procure cotton bales that can be approved by exchanges, and this should reduce the incidence of defaulted contracts. Defaults have become a real possibility at times of price volatility and perceived shortages, and also when hold-ups on grading have occurred. Under the old rules, the latter impediment shot to prominence in 2015. That year Glencore failed to deliver three million pounds of cotton to Noble Agri and actually defaulted on a futures contract.[11] As reported by Greg Meyer, a former Cargill trader claimed on record that this was the first cotton futures default known to him. Before Glencore entered arbitration with Noble on the matter, the defaulting party cited in its defense the fact that it was unable to get its cotton graded before the contract with Noble expired. The new futures benchmark and associated rule changes, when implemented, would in theory prevent the recurrence of similar defaults.

But in the absence of serious position limits this reform might enable greater speculation by the back door. Barriers to entry for physical traders to speculate in the global benchmark have now been significantly reduced. Smaller Asian traders that do not move US-origin cotton, for instance, might now have more incentives to financialize their trading businesses than ever before. If they aim to keep up with the average rates of return in the global cotton-trading business, they might now have to turn to the short-term profit-making strategies that the major traders have employed. Absent more rigorous regulation, the new global benchmark could become a lightning rod for financialization, and not simply the vehicle for enhanced risk management that supporters such as the American Cotton Shippers Association have professed it to be.

This is especially worrisome, given that at least one leading individual cotton trader has challenged the *moral legitimacy* and legality of the trading strategies employed by the world's top cotton merchant. Mark Allen, the former head of cotton at Glencore, filed a lawsuit against Louis Dreyfus Commodities in a personal capacity in the aftermath of the 2011 cotton price spike.[12] By March of that year, cotton benchmark prices had rocketed past US$2.20 per pound, a level that far exceeded the typical 60–80 cents per pound.[13] In his claim, Allen directly accused Dreyfus of manipulating the market. As the spot price for physical bales declined later in 2011, he alleged that Dreyfus had attempted to inflate benchmark cotton prices for future delivery. Allen claimed that Dreyfus reaped the benefits of a strategy to build a massively bullish or long position in cotton futures directly in May 2011 and again that July. He further asserted that the Dreyfus cotton unit Allenberg

Cotton had turned down traders on the other side of those contracts that had offered to sell it much cheaper cotton in the physical market before the contracts came due.

Glencore and Cargill were the biggest players on the other side of the Dreyfus futures contracts. They had been bearish about futures prices in 2011 and, having taken short or net selling positions, were obliged to deliver cotton to Dreyfus as the May and July 2011 futures contracts expired.[14] Over those months, Dreyfus took delivery of more than half a million bales of cotton. And the controversy stems from how it did so. It would have been cheaper for Glencore and Cargill to deliver any bales that were then available on the physical market. But after their attempts to deliver were allegedly rebuffed and the contracts came due they were forced to abide by the old ICE US-only bales delivery and grading process. As such, they had to pay hefty premiums to procure US bales that met the tough ICE standard, even though cheaper bales were readily available. Prices of physical bales in the spot market had started to collapse after many mills across Asia cancelled their orders with cotton merchants. Earlier in 2011, spinners in need of cotton had entered into high-priced supply deals for future delivery in the context of rising prices. And as firms walked away from those expensive deals, the spot cotton price dropped. Futures prices also fell, but not by as much as they would have in the absence of the allegedly manipulative Dreyfus bet.

So, as prices returned to earth, the world's biggest cotton trader seemed to be the only entity to have emerged relatively unscathed from the futures market chaos. Other merchants, including Noble and Olam, booked big losses over the subsequent months. And in the aftermath many traders accused Dreyfus of having engineered the biggest cotton "squeeze" in history. Some simultaneously blamed the Commodity Futures Trading Commission (CFTC) for failing to diligently enforce its own rules or monitor the market. But, years later, while the lawsuit was still ongoing, the CFTC issued its own report that portrayed the event to have been entirely above board.[15]

And regardless of whether profit-hungry machinations or simple trader stupidity was at the root of the market madness, there are at least two certainties. The first and less consequential truth is that spinners were not at the center of this mess. When many reneged on deals with merchants on the downside, the memories of what had happened to them on the upside were still fresh in their minds. Mills across Asia had been burned by cotton merchants as prices rose in 2010. At that time, in the context of rapidly rising prices, several merchants prominently walked away from the supply deals they had struck with mills when cotton prices had been lower. The subsequent retaliation could not be justified on moral grounds as it undercut the sanctity of contracts and reduced business trust. But this episode of immorality pales in comparison with the dark realities that the dirty 2011 financial drama imposed on small uninsured cotton farmers around the globe. As benchmark prices for cotton lint rose in late 2010 and into 2011, many smallholders failed to be paid higher prices for seed cotton at the farm gate. And as higher world prices persisted, the prospect of receiving higher seed cotton prices induced many to put more cotton in the ground that year. Those that did had their windfall ambitions frustrated after global prices collapsed and farm-gate prices dropped. And on those uneven grounds, the moral legitimacy of financialized cotton was at its shakiest.

The moral dilemmas in this area remain stark. Margin traders continue to be able to take short positions in cotton futures contracts. While their trading strategies offer no guaranteed successes, it remains possible for hedge funds and day traders with access to margin trading to profit from falls in the cotton price – even if they have zero connection to the physical business and no capacity to deliver physical cotton. Financiers would argue that the participation of these types of speculators enhances the efficiency of the market through facilitating price discovery. In this light, the benefits of better, more accurate prices are seen to trump the potential costs of increased volatility. But the morality of a futures trade where actors unconnected to the field can short futures contracts remains dubious at best. The profits or losses generated on the shorting of just one individual 50,000-pound cotton contract for future delivery can be bigger than the annual returns to a handful of average-sized cotton farms across Africa and South Asia.[16] And traders are not simply shorting one contract, one time. This is a day-in, day-out volume business.

The International Cotton Association, for its part, has been a morality enforcer in one limited area of finance in the aftermath of 2011. It has invested considerable time and effort to sensitize mills in emerging cotton-importing countries such as Vietnam about the need to honor the sanctity of contracts with their suppliers.[17] As an industry association and arbitration service, the ICA has kept its focus on securing cordial relations between suppliers and buyers in the physical market. Few ICA members have had an interest in drawing attention to the fact that hedge funds and day traders – financial intermediaries or interlopers – aim to squeeze outsized profits from this sweat-drenched fiber. They do so on their Bloomberg Terminals simply by pointing and clicking. And I for one, on the grounds of social inclusion, continue to wonder what farmers who swing their hand hoes and pick that fiber by hand would think about the ways that traders electronically manipulate cotton and turn it into virtual, financialized reality.

Physical cotton

Most companies that produce and move physical cotton tend not to emphasize the ways that their operations advance democratic principles or detract from their realization. If firms that operate in formally democratic jurisdictions were pressed to do so, they would likely recognize the importance of democracy to their bottom lines – at least some of the time. In the United States, for instance, many enterprises believed until very recently that business-as-usual depended on robust political support for the maintenance of the cotton-specific provisions of the US Farm Bill. And businesses there applauded the political interventions several US senators from cotton-producing states made in support of those provisions over the years.

Yet the upkeep of that support has had grave implications for the realization of democracy elsewhere. US democracy, in the form of its farm bills, spawned a significant negative externality. At the level of the international community, those lavish bills have yielded a lousy spillover. They have discredited many government measures that could have been employed elsewhere in support of cotton, sustainability, and development. And some of the measures that have been defamed actually underpinned the development of agriculture systems in all presently industrialized countries. Policies associated with those bills that supporters of democracy in developing countries might have drawn upon to make cotton more inclusive are now subject to considerable international oversight.[18] They are also much more difficult to introduce. Put another way, the benefits which firms in one democracy derived from policy are increasingly unavailable to those in other democracies. The companies that propped the US system up necessarily contributed to undercutting the policy tool kits of budding or emerging democracies.

And not all companies conceptualize their businesses as incubators for democracy, or vehicles for directing or capturing the benefits that can be derived from it. As Thomas Bassett has elaborated through rigorous research, in the context of West Africa – a region where many states seek to enhance their democratic credentials – the cotton sector has continued to reflect authoritarian impulses. In particular, oligopsonistic market structures and top-down price-setting mechanisms have entrenched the power that seed cotton buyers wield over farmers.[19] And, in the context of the price madness of 2011, farmers in Burkina Faso who sought to have their voices on low producer prices heard were actively suppressed. Farmer boycotts of cotton and crop destruction were forcibly brought to an end as export businesses reasserted their claim to be the agents at the apex of cotton. In that case, several years before the Blaise Compaoré regime's 2014 demise, Burkina Faso's "democratic" solution to support the industry was to call in the gendarmes.

Companies that aim to directly limit democracy through their actions are less and less a part of the story of cotton today. But they do remain a big part of that story. And anti-democratic behavior in this area is not limited to questionable tax avoidance "planning," or to the illicit provision of financial support to politicians who seek to impose pro-business re-regulation or corporate bailouts. In 2015, as talk of sustainability and accountability continued to infuse social media commentary on cotton, many farmers the world over confronted bad buyers at the market. For instance, in Zimbabwe that year, several unscrupulous firms attempted to induce farmers to "defect" from contracts that they had entered into with other buyers to exclusively sell them their produce.[20] To curtail this instance of classic side-selling, a centralized buying scheme was introduced. And this new system – like many tried before it – could potentially mute the voices of farmers that seek to redress industry shortcomings. As Zimbabwe has now sanctioned and formalized a heightened level of business interaction, moving forward, farmers there who seek to challenge prices or call out corporate misconduct now confront more organized business interests. As such, institutionalized cooperation that aims to enable "cleaner" buying is not a one-way street. It can also enhance buyer networking and facilitate corporate schemes to influence future government interventions. Absent broadly shared industry commitments to transparency and accountability, the resolution of similar situations elsewhere could have similarly troubling implications for participatory governance and the development of democracy.

Corporations that contribute to the *democratic legitimacy* of enterprises linked to cotton are nevertheless a growing segment of the global industry. Cotton Connect, for example, is a firm that claims to offer its industry clients individually tailored social-purpose solutions. It delivers sustainability strategy, advice, and capacity building and, for the companies that are willing to retain its services, can also conduct supply-chain mapping. This firm develops, monitors, and evaluates social investment programs and builds so-called "business cases" for mainstreaming sustainability and social accountability into its clients' objectives. Similarly, to enhance democracy along the supply chain, Cotton Connect can convene stakeholder consultations. Moreover, it also assists companies that aim to meet new standards for cotton that require more democratic forms of producer organization, such as fair-trade certification. This firm, and its emulators and competitors, now strive to make money through enhancing transparency and community engagement along the supply chain. In fact, all of the new standards for cotton – including Better Cotton, organic cotton, Cotton Made in Africa, and fair trade – require participation and transparency. More cotton that meets these standards hits the market each year.

And the heart of the matter is that these new standards are not cut from the same cloth. While stakeholder participation was central to the processes that led to the creation of the Better Cotton "standard," several dimensions of this system were not open to question from the outset. As I have detailed in several peer-reviewed journal articles, the Better Cotton Initiative (BCI) rejected the notion that farmers should be paid a price premium prior to the global consultations on the proposed system.[21] The BCI also did not develop principles or objectives related to serious topics that farmers and their advocates in civil society raised during those consultations. Additionally, concerns over gender inequality and persistently unequal household divisions of labor were obscured. Farmer participation with the BCI now typically takes the limited form of top-down capacity building workshops. These are currently undertaken to advance better cotton and not to identify or redress any concerns farmers raise that are unrelated to the system itself. As a consequence, under this standard, those that encounter persistently low farm-gate prices are left to face the relevant public authorities on their own.

Conversely, other standards, including organic and fair trade, have enshrined the payment of a price premium. While higher prices for organic and fair trade have not always fully compensated producers for the costs of converting to those systems, both have contributed to the emergence or renewal of more democratic approaches to farmer organization. In countries where public regulation is ineffective or captured by businesses that source conventional cotton, these institutions have helped farmers to better work together to develop solutions to common problems. For example, in Tanzania, farmer organizations that grew as a direct result of investments in an organic business have identified solutions to obstacles that are unrelated to organic standards.[22] And community reinvestment programs under fair-trade initiatives or the direction of community-oriented cooperatives have, on the other side of the continent in Senegal, yielded similar democratic dividends.[23]

The BCI continues to profess its complementarity with systems such as organic, but the reality is that the better cotton standard cannot yield the same depth or breadth of sustainability. In addition to its notable democratic and economic limitations, unlike organic, it does not impose a blanket moratorium on the use of carbon-intensive inputs. As such, it does not hold the same potential to stimulate a California-type effect on the global scale. This effect can occur when the uptake of higher standards directly encourages new investments and business development linked to meeting those standards.[24] Through setting the bar low in the interest of rolling out a low-cost universal standard, the BCI undercut its capacity to stimulate green innovation. Organic, on the other hand, directly incentivizes business investment that aims to deliver demonstrably greener inputs and services to farmers and their communities.

So, on the grounds of democracy, it is troubling that the BCI continues to be involved with the development of public relations materials that lump these systems together.[25] Journalistic accounts of the system that puff up its "market-friendly approach" have emphasized just how conducive it is for mainstream business interests. These have lauded just how much more "sustainable" cotton hit the market after BCI-"verified" cotton started moving through supply chains. But such advocacy is underpinned by an unstated implication: that other systems unnecessarily raise the costs of procuring "sustainable" cotton. The fact that better cotton is the cheapest approach to securing sustainability should raise eyebrows. It might come at a lower cost than the alternatives – it was explicitly designed to be cut-rate – but that does not mean that it is the most market-friendly approach.[26] More accurately, it is the friendliest sustainability system for the established merchants and branded global retailers that have invested in its success. Others, such as Hennes & Mauritz (H&M), have hedged their bets and now procure mass volumes of certified organic cottons. As one of the world's leading sources of demand for organic cotton, H&M's procurement strategy has also been incredibly market friendly. In particular, its organic purchases have contributed to generating demand for the produce of the new types of sustainability innovators, green disrupters, and impact investors that keep organic cotton going. So while IKEA, Adidas, and a range of other BCI supporters laud the BCI for delivering market-friendly sustainability solutions, the reality is that this system is friendliest to the bottom lines of the biggest firms least in need of a low-cost boost.

None of this means that firms that engage with the BCI do not enhance their *performance legitimacy* in the area of sustainability relative to firms that continue to produce or source conventional cotton. The credentials of its supporters in this area are certainly superior to those that still strive in 2016 to deliver the cheapest, whitest, strongest and "cleanest" fiber using the old agrochemical package. And it must also be emphasized that the performance legitimacy of business does not solely relate to adherence to sustainability standards or the lack thereof. This dimension of legitimacy can cover every commitment that firms make to their shareholders, creditors, regulators, buyers, suppliers, and consumers. As such, it is exceedingly difficult to map performance legitimacy industry-wide. A thorough analysis of just one individual firm's follow-through on its commitments could easily stretch to the length of this book. Just check out US-labor activist Jeffrey Ballinger's penetrating and extensive tweets on Nike's voluminous failings in this area.

In spite of this evident limitation, a major work of social science is not needed to ascertain the area of greatest political dynamism in the business models and commitments that firms connected to cotton make. If you put to the side the development of new technologies, most of the business practices associated with this commodity business have been relatively static since the nineteenth century. Then, the dynamic policy and market "innovations" associated with actually paying people to tend to and pick cotton were associated with political upheavals. Today, the political action in this industry has shifted to sustainability. It is the dynamic challenge and opportunity that bears the most on the future of cotton and its place in the contemporary global economy.

To be evenhanded, each of the new standards can effectively enhance the sustainability of cotton. While the BCI has never aimed to establish a sustainability "policing" mechanism and has never endeavored to build an alternative market, on its own terms it enjoys a level of *moral legitimacy*. Its backers portray the initiative to be a noble business effort to make cotton work better for those that grow it. That it is the lowest-cost approach available to the firms that support it nonetheless raises a serious question about its morality relative to organic and fair-trade systems. That said, it is not the only new standard that has suffered from moral shortcomings. For instance, as I detail in the Afterword below, in the mid-noughties an alleged confidence trickster targeted investors in a proposed organic-cotton project in Tanzania. Similarly, fair-trade cotton has not radiated moral legitimacy when prices paid to farmers have not kept pace with rises in the world price. The politics of morality in the area of sustainability standards is an issue that is worthy of much more context-specific research and comparative study.

And there are now numerous proprietary and open access tools that could be applied to advance such research. Many have been designed and rolled out with the aim of better measuring the various dimensions of sustainability, such as Textile Exchange's incredible organic-cotton assessment tool. Most professional analysts of this fiber now consider better sustainability performance to be a moral imperative.[27] Even Cotton Incorporated, the crop promotion agency, has taken this terrain very seriously. It rigorously examined the environmental impacts of the fiber and the downstream industry in a Life Cycle Inventory Assessment snapshot report published in 2012.[28] This assessment covered a diverse range of externalities from fiber cultivation, including acidification, global warming, and water overuse.[29] A similar framework or hybrid model could be applied to assess the relative performance and morality of the competing standards systems.

Yet the really big moral legitimacy challenges for cotton pertain to the actions taken by firms that have not as yet engaged with the politics of standards and sustainability. In India, for instance, many companies either organize or rely upon teams of child laborers to harvest cotton.[30] It is estimated that the "nimble fingers" of several hundred thousand young people are employed on cotton farms in that country. And many of the children that toil in cotton fields for their parents or for cash there and elsewhere do so for a good reason. They and their families know that such work can bolster family incomes and can also enable other family members to seek off-farm employment. The persistence of voluntary low-cost or no cost household labor is a moral issue that evades universal prescription. But when labor in cotton fields has been forced or coerced, the shadow of slavery-like conditions has once again fallen over this global resource. In 2015, reports again emerged that more than one million children and adults in Uzbekistan and Turkmenistan were being forced to cultivate cotton against their wills.[31] A coalition of human rights defenders, labor unions, and social enterprises known as the Cotton Campaign has worked to raise global awareness of this issue for nearly a decade. Farmers in those countries must lease land from their governments and meet mandatory cotton-production quotas. In both countries, elite and unscrupulous business interests have continued to prop up these unjust and unfree labor systems. At the time of writing, hoods had recently been paid to rough up human rights and labor campaigners in Uzbekistan, and the office of a leading researcher there had been set ablaze.

So while social exclusion remains palpable in Central Asia, elsewhere corporate efforts to foster social inclusion have reached new heights. Companies as diverse as Marks & Spencer, a BCI supporter, and People Tree, an organic and fair-trade clothing specialist, have made prominent philanthropic interventions in support of the development of more inclusive cotton supply chains.[32] However, not all instances of cotton-company giving can be linked to this noble end. A few buyers in East and southern Africa are known to have erroneously characterized disbursements made to regulators to facilitate shipments and smooth regulatory hurdles as "philanthropy." But most firms that operate globally now publicly propound their more legitimately philanthropic activities, and seek to demonstrate that their credentials on this front are bona fide.

The politics of business countervailing business

The great Canadian-American economist John Kenneth Galbraith popularized a concept that can be employed to map the power relations of business. His concept of "countervailing power" appeared above in the conceptual chapter 2 and is reintroduced here to summarize the ongoing geopolitical complexity of the cotton business. Galbraith was particularly attentive to the actions firms take to "countervail" the power of their buyers and suppliers.[33] He focused not only on what businesses try to do to countervail the power of firms on the other side of the market but also on how firms actively organize to countervail the power of non-market institutions, including labor movements and public authorities. Certainly, Galbraith's approach was not uncontroversial. Some in this business today might have an interest in claiming that analyses that apply his concept could define the politics of cotton so broadly that they would amount to no more than belly-aching and navel gazing. But I don't think I need to remind you that there is a thing called belly-button lint for a reason and that the contributions cotton makes to keeping bellies empty should concern us all. A focus on power and politics might matter to a lot of people whose lives are touched by these afflictions or by the other sufferings that this industry continues to fashion.

First and foremost, big businesses in cotton have organized themselves to countervail the power of their buyers and suppliers. Lint exporters and spinning mills actively pushed back against each other during the last big price swing. Since the commodity price boom of the 2000s and through the ensuing price downturn, merchants have also sought to exert more control over the prices they pay farmers or ginners. Olam, for instance, has aimed to do so through acquiring firms that farm cotton and produce lint. Forward from production, some traders have also introduced new approaches to securing sales. Many of their new plans to control the market for the future are linked to their ongoing participation in the lowest-cost sustainability initiative. These traders recognize that the social and environmental qualities of the fibers they push are now part of the demand equation.[34] Successful cotton marketing, in this view, now depends on business efforts to control consumers in the rapidly changing fiber market. And, while some worry that more expensive approaches to achieving sustainable cotton could empower the polyester boogeyman, upstart new entrants have built new markets for organic and fair trade. Ginners and merchants that move conventional cotton now seek to countervail the nascent power of entrepreneurs that offer the latter two innovative cottons. And the big players in financialized cotton have employed another means to control their costs. They can countervail the power of their physical market buyers to choose other fibers through realizing outsized profits in the futures market. Those that do so successfully can effectively subsidize their real cotton businesses.

And splits in this industry between businesses are not limited to the gulf between those that seek to secure bulk market sales, and others that operate on smaller but rapidly growing scales. While a few major traders do move small volumes of alternative fibers, their business models rest on paying suppliers less and rely more on countervailing the power of consumer and labor advocates. Divides on stakeholder engagement are also readily apparent downstream. In the aftermath of factory fires and the Rana Plaza factory collapse disaster, for example, two rival coalitions of branded retailers emerged.[35] The first, the Bangladesh Accord on Fire and Building Safety, was primarily signed by European-based firms, including H&M, Inditex, Primark, and Tesco. The Accord was co-signed by ten labor unions, and gave the latter the power to challenge corporate signatories if they did not follow through on their commitments. Conversely, the coalition of retailers that backed the Alliance for Bangladesh Worker Safety did not co-sign with any labor or consumer groups that could hold members of the Alliance to account.

Beyond Bangladesh, spinners and textile manufacturers in other emerging investment destinations have actively employed their old strategies to countervail the power of their workers to raise costs. As Chinese workers have pushed for higher wages, places such as Cambodia have stood out as new low-cost frontiers. When workers in Phnom Penh protested low wages and scary factory conditions in 2014, cotton buyers there endorsed a wide-ranging government crackdown on workers.[36] Alternative businesses, foundations, campaigns, and a range of new standards seek to countervail the power of business to employ this murderously archaic response and others like it. The Fair Wear Foundation, the Ethical Trading Initiative, the Clean Clothes Campaign, the Global Organic Textile Standard, and the Fairtrade Textile Standard are respectively allied against the industry's twentieth-century business logic and practices associated with securing lowest-cost location. Jeff Ballinger, for his part, remains cautiously optimistic about this movement. He emphasizes the need for activists to go beyond the fleeting feel-goodery of social media interactions and argues that proven strategies, such as boycotts and divestment, will be needed to eradicate sweatshops. The successes and failings of civil society and alternative businesses on this front should be the subject of many more books dedicated to the transformation of cotton and other fibers.

Yet it is worth reiterating that all corporate efforts to countervail suppliers or buyers ultimately remain subject to government politics and policy and also to state-based geopolitics. For instance, when commodity merchants are implicated in export surges that fuel currency appreciation – the so-called "Dutch disease" – they do not only face in-house cost pressures to remain competitive. They also confront heightened political calls to do more for industrial development. And that area of geopolitics is presently fraught. As the Chinese government seeks to prevent the out-migration of its stock of mills, Europe has assisted the African quest to add more value to cotton. Firms that buy and sell cotton around the world are also subject to heightened cost pressures that stem from efforts to curtail subsidies.[37] The power of business can be limited. But it still reigns supreme in many contexts, including in the areas of finance and sustainability standards development. Alternative firms seek to countervail corporate power in both areas, and it is in those contests that the future politics of cotton and its place in the global economy will be forged.

CHAPTER SIX

Beyond the Dirty White Stuff

Politics rules. And politics especially rules the economics and economies of cotton. Everywhere and always. Whether it has been on the land or in company hands, cotton has reliably nurtured and been subject to political visions. It has been a core raw material in efforts to build industrial utopias and has been a key ingredient in idealistic plans for agricultural transformation. And at the very least cotton has been a convenient way for countries to bolster their tax and export revenues – except in places where colonial planners politically imposed inordinate reliance on the sale of cotton to faraway lands. There and elsewhere, the politics of violence has been omnipresent. Fertile soils and immense seas have been violated in the name of cotton. People have been violated where and when they have been forced to cultivate cotton. And the violence done to laborers in cotton mills the world over continues to stain the clothes of fast-fashion capitalism in the present day.

Yet none of the political dirt detailed in this book is inherent to the fiber itself. The world of cotton is in no way foreordained to be nasty or brutish. This global resource is also by no means a permanent fixture of world economic activity or geopolitics. It is not a sacred cow. Today's globalized emperors – the so-called ultra-high-net-worth individuals – could assuredly be clothed in fabrics spun from different fibers. Perhaps even invisible ones. But those with vested interests in cotton reliably seek to snuff out attempts to conjure the political will to do so. Empowered stakeholders have sought to control conversations about doing cotton differently, and to avert "dangerous" questions about whether or not humanity should continue to engage with cotton at all. And it is by no means clear that pointed questions about our common future with cotton would endanger anything more than the pocketbooks of those that have grown accustomed to dining out on the rural toil of millions.

Reforming the cotton order to preserve and reproduce that order is now big business. It is the global industry's overarching political priority. And those that propose alternative approaches to sustainable cotton that are more robust in scope, such as organic certifications, now face a highly organized transnational political machine. Industry supported definitions of "better" cotton can in no way be equated with actual best practices for people and the planet. In my view, those might be rooted in agroecology principles and organic methods applied in context specific and contingent manners.[1] Greater reliance on biological nitrogen fixation and renewable energy can reduce the carbon footprint of cotton significantly and can also generate new sources of rural non-farm employment. Similarly, when farming neighbors work together or pool their on-farm resources to recycle nutrients and conserve soils, water, and energy, they necessarily rely less on petrochemicals.[2] Moreover, approaches to cotton that introduce complementary trap or fertilizer crops tend to enhance ecological relationships and to create new income streams and markets. And to be clear, there are simply no blanket, universally valid prescriptions or silver bullets to be found in the conventional industry's attempts to govern itself or to "capture" regulation or regulators and ward off real change.

The geopolitical power contest to control the future of cotton is ongoing. The Better Cotton Initiative now has a consumer-facing logo, and the commitments that firms such as IKEA have made to solely sourcing "better" cotton have been trumpeted by the World Wide Fund for Nature. The latter organization originated and distributed the idea of "better" cotton in partnership with industry and has continued to take delivery of cash from branded retailers, including Marks & Spencer, to pursue sustainability research.[3] But as the market for garments containing certified organic cotton that have met the Global Organic Textile Standard (GOTS) has hotted up – the number of GOTS-certified mills rose by 22 percent in 2014 – civil society groups and businesses that went all in on "better" cotton now face an ominously political question. Do they want to be associated with rigorous sustainability certifications, or with the vague principles, capacity building, and looser rules that experts have analyzed and equated with "sustainability-lite"?

If we want to see cotton enveloped in a different kind of politics, it will require new ways of thinking about the fiber. First and foremost, fundamentalist belief in the benefits of cotton is not the order of the day. Cotton might be the right choice for the future. And it might not be. It depends. Only independent research can offer informed answers on whether or not cotton should continue to be the fabric of our lives. And more research on the social and ecological footprints of demonstrably different approaches to cultivating cotton in particular places needs to be conducted and compiled.[4] Thinking differently is also a political imperative when it comes to international trade and sustainability. The global market value of organic cotton – before the ecological benefits are factored in – is now worth billions of dollars more per year than the total annual amount spent by all governments on subsidies in support of their cotton sectors.[5] Over the past ten years, while trade ministries have obsessed over trade-distorting cotton subsidies at the WTO, the explosive growth of organic has silently taken the globe by storm. The earth continues to pay dearly when policy makers overlook or obscure organic internationally and at home. However, in spite of rapid demand and volume growth, organic cotton still makes up a very small percentage of the total global value of the world cotton order. Even so, alternative voices seem to be on the cusp of pushing it further into the mainstream.

Yet defenders of the status quo continue to issue ad hominem or personal attacks on those that dare to challenge blind faith in the conventional commodity. A healthy skepticism, or at the very least an agnostic stance on the pros and cons of this global resource, are the minimum requirements for civil political discourse on the future of cotton. But several industry insiders continue to deride the socially or ecologically concerned. At least one former high-ranking international official has publicly claimed that critics are actively undermining the bulk market industry through engaging in "demonization." The faithful, in the absence of rigorous research, also call out their critics through reverting to intellectual "cherry picking." They search for minor holes in arguments that they do not like, such as the presentation of potentially inaccurate statistics. Just ask Neil Young. When the Canadian rock star called for a boycott of non-organic cotton, the conventional industry's rejoinder was to attack some of the statistics in Young's presentation and not the substance of his argument.[6] And the cotton-consuming public can rest assured that "gotcha-type" defenses of the industry will not yield substantively better cotton.

More intellectual pluralism – the fancy way of saying enhanced openness to learning – will consequently help to clean up the dirtier aspects of this business. And if greater receptiveness to dialogue takes hold, new ways of knowing about the merits and demerits of cotton will be required. The danger that a big research push could degenerate into an elaborate drive for the same old kinds of data is nevertheless considerable. For example, new research dollars could be directed toward narrow projects that aimed to identify the specific contributions that each formal aspect of the industry makes to GDP growth in the countries where cotton is cultivated. That unfortunate eventuality would tie any new data linked to that hypothetical research to the sinking ship of GDP.

As former New Zealand finance minister Marilyn Waring has been arguing for decades, GDP is not a measure of genuine progress.[7] When a cotton farmer drinks non-potable water from a discarded pesticide bottle, falls ill, and seeks to visit a hospital, GDP goes up. Fuel has to be purchased, and a driver, a doctor, and a pharmacist presumably need to be paid. Similarly, GDP also goes up when tractors and pickers are purchased in bulk and forests are felled in the service of developing new large-scale cotton operations. Sales agents receive their commissions, and tree-cutters, removers, and "pest eliminators" can bank the proceeds from their temporary employment. The social and ecological value of the land for people, for wildlife, and for the atmosphere before clearance are not subtracted from the national accounts. As such, a more pluralist and forward-looking industry would need to embrace the lesson that it makes little sense to tie the quest for more relevant data in the twenty-first century to inadequate twentieth-century measures of "progress." And they would be in good company if they did so. The persuasively comprehensive report of the Commission on the Measurement of Economic Performance and Social Progress, chaired by Joseph Stiglitz with the assistance of

Amartya Sen, could serve as a guide.[8] True cost accounting demands that we unhook our statistical knowledge from an overly economistic and unidirectional focus on growth.

Unfortunately, the conventional cotton industry's enduring culture of fear could impede or preclude its genuine engagement with new ways of knowing or doing. Fibers that can easily be substituted for cotton and readily cut into its share of the global fiber market, including hemp, linen, silk, wool, and polyester, tend to keep the players with the most skin in the cotton game up late at night. Vested interests would consider any first-mover attempt to seriously assess the social and ecological costs and benefits of conventional cotton and their alternatives, and apply this new knowledge, to be too risky. In particular, a rethink involving new kinds of data and the widespread application of alternative innovative production technologies might contribute to raising the world price of cotton relative to other fibers. And if prices did come to reflect the true costs of cotton done more sustainably, from an ultra-conservative industry perspective, there are no guarantees that the textile and garment sector would not simply opt for ostensibly cheaper fibers, even if they were petrochemical by-products.[9] The route out of this looming quandary leads directly toward enhanced international or global oversight and regulation of the global fiber market. And that is an unpalatable outcome that the full bellies currently at the pinnacle of this financialized bulk commodity business would assuredly not be able to stomach. Unless they could profit from it.

And the reality is that new ways of financing cotton are already yielding serious social and ecological benefits. Stock and bond issuance, bridge loans, credit lines, trade finance, and the current suite of permissible risk-management tools keep the conventional business going day in, day out. If cotton is truly to be done better, venture capital, crowdfunding, philanthropic injections, Islamic finance, and direct giving will need to take higher places in the industry's list of go-to finance options. The latter have each played key roles in the development and expansion of alternative approaches to cotton.

Venture capital, for instance, has helped spinners who want to work with organic cotton to execute viable business plans that might have otherwise gone unfunded. The venture capital arms of several systemically important financial institutions have also permitted up-and-coming organic operators to expand and reinvest in storage. Crowdfunding, for its part, has been tapped to advance the mission of sustainability and consumer-awareness advocates, such as the Ethical Fashion Forum. This transnational network supports the uptake of more sustainable practices in the fashion industry, and facilitates cross-border sustainability-oriented collaborations. In late 2015, the founders of the Forum turned to the British funding portal Crowdcube to seek investors in Mysource, a new tech platform designed to match fashion professionals with more ethical and sustainable suppliers.

Where beating the curve or maximizing shareholder value over quarterly time horizons has been shunned, and longer-term horizons embraced, the business of doing cotton better has become more established. Take for example the philanthropic investment that enabled bioRe Tanzania, an organic cotton operator. It took the better part of a decade for bioRe to turn its first profit. And the social and ecological innovations on that small operation have had an enduring demonstration effect. Neighboring communities have seen that organic cotton can deliver financial, social, and ecological rewards, and the project has now spawned numerous emulators. Sharia-compliant sovereign and corporate debt instruments that are structured in ways that prohibit debtors from being charged interest by their creditors, such as sukuk, could feasibly enhance the capacity of other organic start-ups to deliver similar results. Conditional cash transfers to cotton cultivators could also be rolled out to stimulate the uptake of agroecology principles and on-farm technologies. Moreover, if the big commodity traders that move financial and physical cotton devoted more than 1 percent of their global profits to giving, as many purveyors of certified organic goods now do, governments might not have to foot as much of the bill for transformation. The bottom line here is that even if industry insiders remain fearful of raising business costs, financiers increasingly recognize that there is profit to be found in giving firms the capacity to innovate to build better businesses.

But a cringe-inducing series of political questions continues to be leveled at those that support sustainable development and agricultural change. Simply put, isn't change just too much work? Wouldn't the greater uptake of agroecology and conversion to organic be too labor intensive? And who actually wants to do all of that hard work anyway? Aren't people unwilling to stay on their farms? Furthermore, if incentives shift and keep people on the land, are we not condemning rural people connected to cotton to relatively impoverished futures? The ideological noise on this front is palpable and fundamentally rooted in the old dogma of agricultural economics. Where and when organic is dismissed as "romantic," and fair-trade initiatives are thrown under the bus, obsolete twentieth-century understandings of productivity and efficiency tend to animate the politicking.[10] The knee-jerk reactions of researchers and academics who have spent their entire working lives building abstract theories linked to the maximization of output per unit of input are to a certain extent understandable. But it is unforgivable that empowered scholars who remain wedded to economic models that environmental and technological change have rendered redundant continue to publicly decry innovation.

As artificial intelligence, automation, and robotics make employment more precarious the world over, and talk of guaranteed annual incomes spreads, there is little doubt that regular people are going to need more opportunities to apply their skills. Or to learn new ones, to advance themselves, build communities, and contribute to bringing human activities into better balance with ecosystems. Mass scale employment in factories is a diminishing opportunity. Similarly, more mechanized monoculture will only compound the looming labor glut. As such, the only reasonable contemporary question to ask about efficiency and productivity in agriculture generally, and in cotton specifically, is straightforward: productivity and efficiency of what, for whom?

If alternative approaches to cotton enable more people to be employed, and are more efficient in terms of ecosystem interrelationships, they are by a twenty-first century definition more productive and efficient. The numbers of start-ups that organize smallholder cotton production and that supply or sell new kinds of inputs or crops are growing. And governments have not had to employ the old and discredited heavy-handed techniques of agrarian transformation, such as forced resettlement, to induce new entrants. Rather, many new businesses that adhere to or depend upon alternative approaches to cotton are convinced that cotton zones can become fertile terrains for entrepreneurship.

States nonetheless have serious roles to play. Even if they are ultimately unable to work together collectively or effectively to better regulate the global industry, they can nonetheless incentivize transformation through crafting targeted policies at home. Public research and development funds for cotton can be aligned with the imperatives of social and environmental innovation and commercialization. Investment proposals for large-scale land leases that are light on employment generation and conservation can be knocked back. Social and environmental "impact" investors can be welcomed and nurtured. And public campaigns can be launched to sensitize potential entrepreneurs, job seekers, and consumers about the individual, community, national, and global benefits to be had from alternative approaches. Any of these possible courses of action could dovetail with the efforts businesses and nongovernmental organizations are making to countervail the old conventional wisdom.

Retrograde claims that economic forces will necessarily preclude the transition to new rules of the game for cotton are notable solely for their convenience to the interests of the fiber establishment. Cotton barons and traders and their creditors, agents, suppliers, and buyers have paid increasing lip service to the need for change. The specter of fiber substitution nevertheless continues to haunt the halls at industry events. The faithful fear it most when change agents demonstrate their growing market and political power. And it is in that countervailing power that we can find hope that the seeds of a better future with – or without – cotton are being sown. The politics of this global resource are on display every day all around us, whether we are between the sheets or on the streets. Those that have cared enough to delve more deeply into the fiber-to-fashion story have enabled social and environmental change through activism and advocacy. And viable investments that have fostered new kinds of consumer choices have flowed from such activism.

The final question, then, is not whether we should be prepared to pay more for cotton to better reflect its true costs. The question is when we will be asked to pay more for cotton. Nobody that wishes to countervail the power of big cotton seeks to turn standard bulk-market cotton into an exclusive, high-end product. The luxury cotton product niche will be with us as long as capitalism continues kicking. But the keys to a fluffier future can be found in efforts to control the middle of this global resource differently and to transform its commodity politics.

Afterword: A Learner in the World Cotton Order

So how exactly did a guy from a place where snow tends to be the only white stuff that covers the fields each year end up writing this book? The short answer is that cotton was about the farthest thing from my mind as I grew up on the shores of Georgian Bay, Ontario. Swimming, skiing, and canoeing were my biggest after-school and summer pursuits. As such, when I had to make decisions about new outdoor clothes, breathability and waterproofness were always at the top of my list. Most of the time, cotton was consequently ruled out. I did buy a few preppy branded casual shirts and jeans during my high-school years and had a pair or two of canvas Converse. But the fabrics of my life into my twenties were silk, wool, and Gore-Tex. And thankfully cotton didn't really make it into my diet either. As far as I know, crystallized cottonseed oil was never on the menu at home.

After I left Queen's University in the middle of my studies on a letter of permission to pursue love and fortune down under for a year, my obsession with non-cotton outerwear persisted. Suffice it to say that Aussie fashion was polyester-heavy at the dawn of the new millennium. Upon returning to Canada, to fund my ongoing education I continued to plant trees on a piecework basis for reforestation companies in Northern Ontario. We received 7–9 cents per tree, some of which I reinvested in protective nylon pants, neoprene socks, and rubber boots.

So when I found myself back at Queen's taking a course entitled "critical perspectives on contemporary capitalism" during the fall term of 2000, I was strutting around in a Gore-Tex jacket. It wasn't too long after I cracked Naomi Klein's bestseller *No Logo: Taking Aim at the Brand Bullies* that I picked up a permanent marker and proceeded to black out the prominent logo on the right sleeve of that jacket.[1] Something in Klein's critique of factory-work conditions, branding, marketing, and profiteering resonated in my mind. I reread William Greider's *One World, Ready or Not: The Manic Logic of Global Capitalism* and decided that political economy was where it was at for me.[2]

In hindsight, my timing was particularly apt. Political economy analyses of corporate-driven globalization were then informing a worldwide movement that had shot to prominence the previous year. While I had been wearing a wool suit and working at a call center in Sydney, this movement burst into public view at the Seattle Ministerial meeting of the WTO. In the wake of the so-called "Battle in Seattle" and firm in my new knowledge, I found this movement's critiques of the increasing corporate dominance of international relations to be particularly convincing. As such, I became involved with the alter-globalization movement. Our little group – the Queen's Coalition against Corporate Globalization – stood up against the exclusivity deal Coca-Cola had struck with our university. We also challenged pharmaceutical firms that had a presence on campus. At that time, several continued to employ their drug patents to fatten their profits at the expense of the poor and sick in Africa. And in April 2001 we made campus headlines again. The coalition contributed to a campaign that encouraged the University Senate to excuse students of all political stripes from their exams so that they could attend the third "Summit of the Americas" in Quebec City.

And when we were ultimately able to defer our exams and make our voices heard in Quebec, I had two life-altering experiences on the streets. I learned firsthand what it was like to be on the receiving end of tear gas and, at the start of the official protest march, I also felt for the first time the true power of a discarded banana peel as I literally and unexpectedly slid down the road.

So it should come as no surprise that later that year when Klein's analysis hit the cover of the *Economist* magazine that I was sitting down to read the great Brazilian educator Paolo Freire. His famous book *Pedagogy of the Oppressed* offered its readers a perspective on how it might be possible for both the oppressed and their oppressors to liberate themselves through education.[3] And it was on that very sad September day in 2001 that a group of people who were obviously under the influence of a very different and fundamentally dangerous approach to learning carried out their awful attacks on the United States. I subsequently endeavored to better understand the various ways that oppressed people have sought to liberate themselves from economic injustice, and the non-violent means that they have utilized to do so internationally.

This interest led me to develop a major research focus on the Third World project to transform the world economy. I happened on this topic as part of my research for a study group on the WTO's "Development" Agenda convened by Daniel Drache and Marc Froese at York University. After Daniel reached out and sucked me in on my first day of grad school, I quickly realized that I needed to up my understanding of economics and its politics. I also started to notice that my clothes had become more than a bit tatty. So while I immersed myself in Toronto's used-clothing scene, I boned up on the dusty declarations and programs of action that developing countries issued collectively during the heady days of commodity or resource power. But even after Drache cracked his editorial whip and shepherded my project through the finish line, I hadn't yet come to cotton. That took a broken cell phone and some good timing.

In September 2003, a WTO Ministerial meeting was held at Cancun. This was the first Ministerial held after the "Development" Agenda was agreed at Doha in the aftermath of 9/11 and the Battle in Seattle two years earlier. Several initial accounts of the Cancun meeting suggested that a new North–South divide had emerged over efforts to make cotton work better for development.[4] And that message was in the back of my mind when I defended my Master's on the twenty-first-century relevance of the Third World project. Stopping my subway trip home from the defense at a mall, I struck up a conversation with a phone company sales rep. He told me he had recently completed graduate-level training in politics. I told him I had just defended a project on commodities. And he told me about his cotton-farming aunt and uncle in Uganda, and their many struggles to make cotton pay.

I was hooked. Here was a commodity that was associated with high-level global politics. Cotton seemed to have become a make-or-break issue for African countries at the WTO. And international rhetoric on this topic was heated. So I pitched the idea of pursuing a PhD on globalization and cotton to William D. Coleman and Robert O'Brien at McMaster University. As I planted trees on Vancouver Island and waited to hear back regarding my application, I planned an initial trip to East Africa. Knowing little about the context beyond what I had read, I wanted to test myself to see if I would have the physical and mental strength necessary to conduct real research on cotton in Africa. I read Ngugi wa Thiong'o, Colin Leys, Chinua Achebe, M.G. Vassanji, Ousmane Sembene, John le Carré,

Bruce Berman, and Brian Cooksey. And McMaster said yes to me in the context of William Coleman's broader research success. He had just landed significant research funds to launch a transnational collaborative project on Globalization and Autonomy.

To be clear, I have not presented this background simply to emphasize the point that my clothes eventually got better: I have not intended to be self-aggrandizing whatsoever. Rather, this elaboration has aimed to impart just how remote I was from cotton before this commodity became wedded to my career. I remain a white guy from Ontario. I also happen to have been granted tenure at a research-intensive, comprehensive university in that province based partly upon my studies of cotton in Africa. Consequently, it is incumbent upon me always to endeavor to situate myself in relation to my area of study. After all, it is possible that an individual from any walk of life in the countries where I have conducted research could raise questions about what I have been up to at any time. For instance, someone could legitimately ask why anyone from Ontario thinks they have any right to write or speak with authority on power and poverty in Africa. They could also challenge my credentials or question my impartiality.

The latter questions have prominently been leveled at another development studies professor whose work focuses on the problems associated with the production of reliable statistical information on Africa. Over the past several years, Morten Jerven's publications have elicited unhelpful off-the-cuff reactions from those with vested interests in the statistical status quo.[5] Government officials in several African countries have raised questions in public about the veracity of his research. Some have even associated Jerven – without presenting any evidence to back their claims – with a nebulous transnational conspiracy that aims to discredit the notion that Africa is "rising" and on the road to political economic transformation.

To date, my publications and findings on cotton have not been subject to unwanted or unwarranted attention or criticism. Valid questions about my legitimacy as a researcher studying an issue of importance to Africa have also not been raised. So why might this be the case? Speculatively, it could be that the seemingly innocuous nature of cotton has helped me out a bit. After all, in many African places, cotton is only a small sub-sector of much larger agricultural sectors. Documenting and calling out the wrongs committed by players of all types in the name of cotton could be seen to be less consequential for businesses in Africa than research that questions the underpinnings of economic knowledge on that context.

But there could be another reason for why my voice as a western expert has not yet been dragged into the court of public opinion. From the outset, I have been publicly attentive to my situation in relation to the research. When I commenced my PhD fieldwork on globalization, cotton, and poverty in Africa, I openly discussed the challenges I faced as I learned about how to appropriately conduct research in Tanzania and Senegal. I wrote a blog – *Cottonundrum* – that often focused on situations related to research ethics.[6] Evident on those pages are my initial musings and thoughts about a diverse range of complications that I navigated for the first time. For instance, I was open about my steep learning curve when it came to local research cultures, and about my reliance on house help and the goodwill of people to entertain yet another white researcher. And I wrote about my relatively empowered situation in relation to locally based researchers and what I thought I could do to add value as a foreign researcher.

I am not sure whether I ultimately met my aspirations on

the latter. But I endeavored to do so. My approach to learning about the forces and factors that have impoverished Africa's cotton producers and to assessing how and in what ways new actors and approaches were altering those historic relations was bottom up. What that means is that I embraced an inductive rather than deductive approach to research. I did not enter my research sites with a theory in mind that I wanted to test in the sole interest of improving that theory. Instead, I embraced a question or problem-based approach to learning about cotton, poverty, and change. And the problem was straightforward. I asked how and in what ways companies and non-profits were altering the realities of cotton on the ground.

To arrive at answers to my question, I relied upon stories that stakeholders in cotton imparted to me during extended interviews and focus groups. I also collected and analyzed documents related to cotton that companies, ministries, and non-profits authored or provided. And I observed and participated in the life of the development and agricultural research communities in Tanzania and Senegal. But I did not conduct any surveys. The potential socioeconomic costs and risks associated with the execution of surveys or bona fide "participatory action research" in marginalized, remote communities that depended on cotton simply seemed too high to me. I had limited means. I also did not want to open individuals that engaged with me up to potential retribution. Moreover, surveys could have landed me in trouble with local authorities had I subsequently published unauthorized statistical data. And action research could have yielded the bogus charge that I simply aimed to incite rural unrest. Consequently, my field research relied for the most part on an anthropological tool kit, supplemented with a political science knack for securing and conducting elite-level interviews. The methodological road not taken was nonetheless a limitation, and it is a road that I am committed to taking in the future.

The flow of information that has stemmed from my research on cotton in Africa has necessarily been associated with the exercise of power in the world cotton order. In writing about cotton and change in developing countries I have exercised power, and I have also challenged the power that others wield. On the former, I have in some cases spoken up about and written on behalf of those whose voices and viewpoints on cotton are not often heard. Giving voice to those who have been economically marginalized or dispossessed of their land in the name of cotton is one thing. Doing so in a manner that does not further marginalize or dispossess people is quite another. For instance, I knew firsthand from my research that targeted efforts to transcend one source of oppression in Africa – such as the transboundary spillover from cotton support policies in China or the United States – could unwittingly enable other stark problems to be overlooked. In my estimation, the framing of African cotton as primarily a trade issue at the WTO had done exactly that.[7] In playing up this angle, governments and nongovernmental organizations in some cases obscured other consequential challenges, such as financialization and the emerging power of corporate self-regulation.

As I saw it, my responsibility as a researcher was to capture and re-present cotton problems in all of their diversity. I figured this approach would enable me to offer a more comprehensive understanding of the politics of change, and the prospects for poverty reduction, than the dominant trade-centric narrative. And I aimed never to let myself forget that many others with stakes in African cotton did not have the privilege of thinking or writing about cotton. As I pecked away at my keyboard, they had to work with it, or try to make it work better, day in and day out. I also knew that how I wrote about cotton could advance or undercut the interests of all kinds of different people with stakes in cotton. And I was perhaps too aware that none of these considerations mattered if my research did not get published.

After my dissertation was successfully defended, the only public record of my research was my blog. Having simultaneously landed a tenure-track job, immediately thereafter I began to piece together potential scholarly outputs. I pitched a book proposal and wrote a conference paper on an aspect of my research linked to corporate social responsibility and poverty. In writing that paper, as per my convictions above, I called out the conduct of several institutions and individuals connected to cotton in East and West Africa. While a heavily revised version ultimately became a chapter in my book *Governing Cotton*, I pulled no punches in that initial draft. It was written to speak directly to the evidence I had gathered on the contributions that businesses were or – more often – were not making to poverty reduction. But my "keep it real" approach to scholarship did not seem to attract the attention of many attendees at the British International Studies Association (BISA) annual meeting. The draft paper as presented did not speak to theory directly. And the theorists definitely did not seek me out.

After I returned to Canada, feeling a little dejected about the prospects for mobilizing my knowledge, I set about preparing to deliver my first lecture-based course. And after subjecting my students to a few truly awful first attempts, days later, I received an email that forever altered my understanding of scholarly impact. From that email, I learned that BISA had made all of the papers presented at the recent meeting publicly available on its website. In it, a globally recognized expert in high-altitude medicine and a respected mountaineer wrote that he had come across my draft. And he expressed a keen interest in the sections of that draft that pertained to an alleged cotton scam. I knew that the last thing mountaineers would be interested in wearing was cotton. So I read on. And I am so glad that I did. The situation was straightforward: a team of elite mountaineers was then busily preparing for an expedition to K2. But they were encountering problems with the logistics "fixer" they had retained to help them organize their approach to the world's most difficult summit. Trust between team members and their fixer had degenerated to such an extent that the former had taken to scouring the internet to find out more information on their agent's previous activities. And it just so happened that one of the members of the team recalled the fixer's casual offhand claim that he had been involved in "ethical" cotton in East Africa.

The K2 team happened across my draft because I detailed an alleged "ethical" cotton fraud. I did so because that alleged fraud came up in each of the elite-level interviews that I conducted on cotton in Tanzania. The gist of the story my interviewees told me, and that I was able to triangulate, was that in the mid-2000s numerous local investors had fronted a very convincing "social entrepreneur" a lot of money. He claimed to be readying a grand-scale project to convert thousands of farmers to organic cotton techniques in the Handeni District. A seemingly noble cause. Yet after high-level fundraising and launch events at a top hotel and in the community that would host the proposed investment, the project never materialized. And the money subsequently vanished, along with the "impact investor."

In light of this information and also of reports the K2 team received from investigators in Pakistan that seemed to confirm their organizational concerns, the mountaineers reportedly abandoned their planned expedition. The whole affair subsequently spilled into the public domain after an unrelated third-party wrote a post on a blog about K2, mountaineering, organic cotton, and the alleged confidence man. To this day, individuals around the world who claim to have subsequently been relieved of their funds or jobs or both continue to post comments on that blog, as do others that profess his innocence or seek to defend his reputation.[8]

I had not anticipated that I would be able to exercise power in this small way. Perhaps the effort I made to give voice to the concerns people expressed to me in confidence did some good. But that good pales in comparison to the goods that could have potentially been realized through the establishment of a veritable contract-based organic-cotton operation. In the wake of this alleged fraud, in 2013, a Japanese investor secured the rights to 40,000 hectares of land in the same district to set up a commercial-scale cotton farm. Massive commercial farming simply does not hold the same potential to reduce poverty sustainably and ensure food security as Tanzania's successful smallholder organic-cotton model. The latter, while clearly not immune from abuse, has improved and changed many lives.

Having blown a whistle, I stepped back and focused on how I could distill my findings into products that would be acceptable to an academic audience. After all, I needed to secure tenure and I was fairly certain that this type of "community engaged scholarship" did not fall clearly within my department's criteria for promotion. On that front, my book *Governing Cotton: Globalization and Poverty in Africa* was published in 2011.[9] I launched the book in South Africa, Tanzania, and Cameroon and was able to reconnect with some of the people that made my dissertation research possible. And over the next few years my research agenda backed away from cotton a bit. I analyzed the perspectives that contended to frame food security challenges in Central Africa and made five extended research visits to Cameroon with my wife Lauren. That said, I did publish a few articles on cotton in scholarly journals. I also wrote up an op-ed for the *Financial Times* magazine, *This is Africa*, on the ongoing challenges facing African cotton at the WTO.

The latter knowledge product on cotton was nothing special. It was a short and rather direct piece on geopolitics that called out the settlement Brazil and the United States had reached in their long-running trade dispute for what it was. The deal, as I saw it, simply enabled Brazil to receive compensation that was directly linked to augmenting its capacity to export cotton. As such, I argued that the settlement was a raw deal for Africa's cotton exporters. It simply enhanced the capacity of one of their principal competitors. After that piece was out, I focused my energies on the food security politics side of my research agenda and didn't think much more about it. And then one day in April 2015 my phone rang and the WTO Secretariat was on the other end of the line.

Actually, the WTO had emailed me first. But both came as quite a shock. My book had spelled out at length numerous consequential failings at the WTO in the aftermath of the African cotton initiative. And as it turns out, they were not calling to inform me of a pending legal action. Rather, the WTO wanted me to moderate a panel on African cotton.[10] For me, this was a curious request. I am not known to those that know me best as a moderating voice. Having set out to challenge the trade-centric framing of African cotton in 2004 and, having done so in my scholarly outputs, I had to make a choice. If I declined to participate in the event, I could distance myself from a geopolitics that I knew was suboptimal at its core. I continued to believe that there were better ways for Africa's cotton powers to advance their collective interest in reaping a better deal from this fiber. I also wondered if my participation would in some ways sanction or legitimize flawed WTO processes. But if I declined the invitation, I would assuredly cede the ground at the WTO to the status quo. Pondering this, I came to the realization that if I did moderate the panel, I would be able to contribute to framing some of the broader development challenges. Given the latter consideration, it really was no choice at all. I indicated to the Secretariat my willingness to be attentive to the need for "sensitivity" on panels of high-level personalities, and an invitation from WTO Director-General Roberto Azevêdo subsequently arrived in my inbox.

So too did a script. To secure the participation of the trade and economy Ministers of the Cotton Four (C4) countries and of the heads of international organizations, including the International Cotton Association and the International Cotton Advisory Committee, the Secretariat spelled out just what questions I would be posing to panelists. This was a very reasonable thing for them to do, given the need for diplomacy and the reality that there would also be a world banker and a representative of the USAID on the panel. But I did land a prescribed three to five minutes of freedom to introduce and frame the session. As the focus was to be on reducing trade costs in the cotton value chain – perhaps the most trade-centric framing possible – I had to zero my planned remarks in on the business side of cotton. So, despite its evident shortcomings, this topic enabled me to comment on some of the juiciest global politics related to business, and sidestep the inter-state, international issues that had been deemed "off limits." It permitted me to raise questions about the applicability of old definitions of "productivity" and "efficiency" for cotton, given the twenty-first-century imperatives of redressing global environmental change and enhancing global food security.

When I arrived in Geneva with my wife Lauren to moderate the high-level cotton panel at the 5th Global Review of Aid for Trade, I realized what kind of power I had secured in the cotton order. The cost of our modest hotel room for one night was equivalent in value to the earnings that many small Burkinabe households could hope to scratch from cotton over the course of the whole year. As I witnessed the big Mercedes with diplomatic plates rolling up to WTO headquarters to deliver ministers and trade officials into the outstretched arms of the director-general, I could not help but think how remote this whole exercise was from cotton. Yet everyone in attendance seemed to be wearing it.

The panel ultimately went off without a hitch. I was able to get my two cents in on "productivity" and "efficiency" in the African cotton chain, and on how the diversification of viewpoints on what those two terms mean could present the continent with a big opportunity. And when a panelist noted that their organization did not foresee much potential for growth in the organic sector, I quickly assured the audience that I was wearing a cotton suit that was certified to have met the Global Organic Textile Standard (GOTS). Actually, I was clothed head to toe in GOTS-certified organic cotton from People Tree, Arthur & Henry, bgreen apparel, prAna, and Coyuchi. Some might say that you can "feel" the justice in it. I just think it feels better, and I am convinced that it does less harm than the alternatives.

Afterward, I knew that the whole exercise did not change very much in the cotton order. But the very fact that I was invited seemed to indicate that a few powerful figures in that order were willing to think differently about the cotton "problem." At the very least, the experience gave me keen insight into how "insiders" conceptualize development and change in the cotton order. And those insights were invaluable as I subsequently sat down to write this book. Fifteen years earlier, I might have taken a different approach at the WTO. There are, after all, a lot of potential ways in the internet age to create a social media sensation. But as a learner, and someone with an interest in mapping the contours of power, I felt very content to have limited myself to being a participant observer. As an academic, it was the right thing to do. I think.

And so I turn it over to you. What is the right thing for you to do in the cotton order? Are you going to order new ethical undies or socks online when you put this book down or throw it across the room in frustration? How are you going to continue to engage with the reality that you are what you wear? And if you are too cool to care or have a problem with people that do and you also made it this far into this book, I have to say congratulations. Whether you like it or not, you are a participant in the cotton order. And also in many other orders covered in the Resources Series. Perhaps not to the extent that you dine out on your participation, as I and the other authors featured in this series absolutely do. But maybe one day you might. As a cotton consumer, if you go forward thinking like a global citizen, at least you now have some evidence at your fingertips to suggest that you might be on the right side of history. And that might not be such a bad ending for you or for the rest of us here on this little planet.

Notes

1 SPINNING A FIBROUS TALE

1 Jeff Ballinger, "How Civil Society Can Help: Sweatshop Workers as Globalization's Consequence," *Harvard International Review*, Summer 2011: 54–9.
2 Koray Çalişkan, *Market Threads: How Cotton Farmers and Traders Create a Global Economy* (Princeton: Princeton University Press, 2010).
3 Giorgio Riello, *Cotton: The Fabric that Made the Modern World* (Cambridge: Cambridge University Press, 2013).
4 Claire Delpeuch and Antoine Leblois, "Sub-Saharan Cotton Policies in Retrospect," *Development Policy Review* 31(5) (2013): 617–42.
5 Jennifer Clapp, *Food*, 2nd edn (Cambridge: Polity Press, 2016).
6 Peter Ton, *Cotton and Climate Change: Impacts and Options to Mitigate and Adapt* (Geneva: International Trade Centre, 2011).
7 Eric Hazard, *Le Livre Blanc sur le Coton: Négociations Commerciales Internationales et Réduction de la Pauvreté*, 2nd edn (Dakar: ENDA Tiers Monde, 2005).
8 Lucy Hornby, "China Abandons Failed Cotton Stockpiling Program," *Financial Times*, January 20, 2014.
9 Amartya Sen, *Development as Freedom* (New York: Oxford University Press, 1999).

2 THE WORLD COTTON (DIS)ORDER

1 Adam Sneyd, "When Governance Gets Going: Certifying 'Better Cotton' and 'Better Sugarcane,'" *Development and Change* 45(2) (2014): 231–56.

2 Douglas Farnie and David Jeremy (eds), *The Fibre that Changed the World: The Cotton Industry in International Perspective, 1600s–1990s* (Oxford: Oxford University Press, 2004).
3 Jagdish Bhagwati (ed.), *The New International Economic Order: The North–South Debate* (Cambridge, MA: MIT Press, 1977).
4 Peter Gibbon and Stefano Ponte, *Trading Down: Africa, Value Chains and the Global Economy* (Philadelphia, PA: Temple University Press, 2005).
5 Robert W. Cox, "Ideologies and the New International Economic Order: Reflections on Some Recent Literature," *International Organization* 33(2) (1979): 257–302.
6 Robert W. Cox, "Social Forces, States and World Orders: Beyond International Relations Theory," *Millennium* 10(2) (1981): 126–55.
7 Craig N. Murphy, *The Emergence of the NIEO Ideology* (Boulder, CO: Westview Press, 1984).
8 ICAC, *Structure of World Cotton Trade: May 2011* (Washington, DC: International Cotton Advisory Committee, 2011).
9 Jennifer Clapp and Doris Fuchs, "Agrifood Corporations, Global Governance, and Sustainability," in J. Clapp and D. Fuchs (eds), *Corporate Power in Global Agrifood Governance* (Cambridge, MA: MIT Press, 2009): 1–25.
10 Christopher May, "Strange Fruit: Susan Strange's Theory of Structural Power in the International Political Economy," *Global Society* 10(2) (1996): 167–89.
11 Kevin Watkins, *Cultivating Poverty: The Impact of US Cotton Subsidies on Africa*, Oxfam Briefing Paper 30 (Oxford: Oxfam International, 2002).
12 Stefano Ponte et al. (eds), *Governing Through Standards: Origins, Drivers, Limitations* (Basingstoke: Palgrave Macmillan, 2011).
13 Harold Witt, Raj Patel, and Matthew Schnurr, "Can the Poor Help GMO Crops? Technology, Representation & Cotton in the Makhathini Flats, South Africa," *Review of African Political Economy* 33(109) (2006): 497–513.
14 William Moseley and Leslie Gray (eds), *Hanging By a Thread: Cotton, Globalization, and Poverty in Africa* (Ohio: Ohio University Press, 2008).
15 Simon Ferrigno and Alfonso Lizarraga, "Components of a Sustainable Cotton Production System: Perspectives from the Organic Cotton Experience," *ICAC Recorder*, March 2009: 13–23.

16 ICAC Expert Panel, *Measuring Sustainability in Cotton Farming Systems: Towards a Guidance Framework* (Washington, DC: ICAC Expert Panel on the Social, Environmental, and Economic Performance of Cotton, 2014).
17 Textile Exchange, *Organic Cotton Market Report 2014* (Lubbock, TX, 2015). http://textile exchange.org/resource-center/media-room/2014-organic-cotton-report.
18 Claire Delpeuch and Anneleen Vandeplas, "Revising the 'Cotton Problem' – A Comparative Analysis of Cotton Reforms in Sub-Saharan Africa," *World Development* 42: 209–21.

3 COTTON ON THE LAND

1 Deborah Bryceson, "Deagrarianization and Rural Employment in Sub-Saharan Africa: A Sectoral Perspective," *World Development* 24 (1996): 97–111.
2 UNCTAD and UNEP, *Organic Agriculture and Food Security in Africa* (New York/Geneva: UNCTAD/UNEP, 2008).
3 World Wildlife Fund, "The Impact of Cotton on Fresh Water Resources and Ecosystems," WWF Background Paper (Gland, Switzerland: WWF, 1999). http://www.assets.panda.org/downloads/impact_long.pdf.
4 D. Clayton Brown, *King Cotton in Modern America: A Cultural, Political and Economic History since 1945* (Jackson, MS: University of Mississippi Press, 2011).
5 USDA, "Cotton: World Markets and Trade," *Foreign Agricultural Service Circular Series* (Washington, DC: USDA, March 2012).
6 Felix G. Baquedano, John Sanders, and Jeffrey Vitale, "Increasing Incomes of Malian Cotton Farmers: Is Elimination of US Subsidies the Only Solution?," *Agricultural Systems* 103(7) (2010): 418–32.
7 Adam Sneyd, *Governing Cotton: Globalization and Poverty in Africa* (Basingstoke: Palgrave Macmillan, 2011).
8 Colin Poulton et al., "Competition and Coordination in Liberalized African Cotton Market Systems," *World Development* 32(3) (2004): 519–36.
9 Frank Mulder, "Counting the Cost at 2600 Litres of Water a T-Shirt," *IPS News*, March 24, 2012.
10 David Tschirley et al., "Institutional Diversity and Performance

in African Cotton Sectors," *Development Policy Review* 28(3) (2010): 295–323.

11 Godwin Masuka, "Agricultural Liberalization, Cotton Markets and Buyers' Relations in Zimbabwe, 2001–2008," *Singapore Journal of Tropical Geography* 34 (2013): 103–19.

12 Matthew Schnurr, "Cotton as Calamitous Commodity: The Politics of Agricultural Failure in Natal and Zululand, 1844–1933," *Canadian Journal of African Studies* 47(1) (2013): 115–32.

13 Anthony Winson, *The Industrial Diet: The Degradation of Food and the Struggle for Healthy Eating* (Vancouver: University of British Columbia Press, 2013).

14 For more on food politics and security in marginalized places that grow cotton and depend upon it, see William G. Moseley, "Neoliberal Policy, Rural Livelihoods and Urban Food Security in West Africa: A Comparative Study of the Gambia, Côte d'Ivoire and Mali," *Proceedings of the National Academy of Sciences of the United States of America* 107(13) (2010): 5774–9.

4 COTTON FOR THE COUNTRY

1 Sven Beckert, *Empire of Cotton: A Global History* (New York: Alfred A. Knopf, 2014).

2 ICAC, *Production and Trade Policies Affecting the Cotton Industry* (Washington, DC: ICAC, 2015). Note that the total global market value of certified organic cotton was estimated to be nearly US$16 billion that year. As such, the value of the subsidies governments offer conventional cotton farmers pales in comparison with the total market value of a small but rapidly growing and ecological consequential segment of the global market.

3 James C. Scott, *Seeing Like a State: How Certain Schemes to Improve the Human Condition Have Failed* (New Haven: Yale University Press, 1999).

4 Gregory Meyer, "Cotton Slides on Bumper US Harvests," *Financial Times*, August 19, 2014.

5 Textile Exchange, *Organic Cotton Market Report 2014* (Lubbock, TX, 2015).

6 "Le Burkina Dit Stop aux OGM de Monsanto," *Jeune Afrique*,

June 1, 2015. http://www.jeuneafrique.com/233742/economie/le-burkina-dit-stop-aux-ogm-de-monsanto/.

7 Brian Cooksey, "Marketing Reform? The Rise and Fall of Agricultural Liberalization in Tanzania," *Development Policy Review* 21(1) (2003): 67–91.

8 Ha-Joon Chang, "Rethinking Public Policy in Agriculture: Lessons from History, Distant and Recent," *Journal of Peasant Studies* 36(3) (2009): 477–515.

9 *Radio France International*, "Benin President Pardons 'Poison-Plot' Businessman Who Fled to France," *RFI*, May 15, 2014. http://www.english.rfi.fr/africa/20140515-benin-president-pardons-poison-plot-businessman-who-fled-france.

10 Gawain Kripke, "As the Moon Follows the Sun, US Cotton Producers are Whining for More Subsidies," *Oxfam America Politics of Poverty Blog*, December 15, 2015. http://politicsofpoverty.oxfamamerica.org/2015/12/as-the-moon-follows-the-sun-us-cotton-producers-are-whining-for-more-subsidies/.

11 Personal communications, High-Level Cotton Panelists, WTO Headquarters, Geneva, July 2, 2015.

12 Alan Isaacman, *Cotton is the Mother of Poverty: Peasants, Work and Rural Struggle in Colonial Mozambique, 1931–1961* (London: Heinemann, 1996).

13 Robert Bates, *Markets and States in Tropical Africa: The Political Basis of Agricultural Policies*, updated edn (Berkeley: University of California Press, 2014).

14 John Baffes, "The 'Cotton Problem,'" *World Bank Research Observer*, April 2005.

15 Christopher Langner and David Yong, "Temasek Drags Olam From Muddy Waters to Winning $1 Billion Loan," *Bloomberg*, November 2, 2015.

16 Robert Hunter Wade, "What Strategies are Viable for Developing Countries Today? The World Trade Organization and the Shrinking of 'Development Space,'" *Review of International Political Economy* 10(4) (2003): 621–44.

5 COTTON IN COMPANY HANDS

1 Jan Aart Scholte, "Civil Society and Governance in the Global Polity," in Morten Ougaard and Richard Higgott (eds), *Towards a Global Polity* (London: Routledge, 2002): 145–65.
2 For an application of their framework to cotton in Africa, see Adam Sneyd, "Governing African Cotton and Timber through CSR: Competition, Legitimacy and Power," *Canadian Journal of Development Studies* 33(2): 143–63.
3 For an extended discussion of the components of financialization, see Jennifer Clapp's accessible chapter in the second edition of her book *Food* (Cambridge: Polity Press, 2016).
4 Oxfam, "Not a Game: Speculation Versus Food Security – Regulating Financial Markets to Grow a Better Future," Issue Briefing, *Oxfam International*, October 2011.
5 Neil Hume, "Louis Dreyfus Commodities Gets 'IPO-Ready,'" *Financial Times*, April 1, 2014.
6 Alex Cobham et al., *Estimating Illicit Flows of Capital via Trade Mispricing: A Forensic Analysis of Data on Switzerland*, Working Paper 350 (Washington, DC: Center for Global Development, 2014).
7 Gregory Meyer, "Push Against Commodities Plan," *Financial Times*, February 10, 2014.
8 Alex Cobham et al., *Estimating Illicit Flows of Capital via Trade Mispricing*.
9 Emiko Terazono and Javier Blas, "Commodities Trading Rule Calls Rejected," *Financial Times*, March 26, 2013.
10 Gregory Meyer, "Global Cotton Futures Contract Clears Hurdle," *Financial Times*, June 17, 2015.
11 Gregory Meyer, "Glencore Cotton Deal Comes Unstitched," *Financial Times*, October 1, 2015.
12 Gregory Meyer, "Court Battle Over Cotton Price Gyrations Centres on Supplies," *Financial Times*, May 21, 2013.
13 Josephine Mason, "Exclusive: Top Cotton Trader Allenberg Loses Second in Command," *Reuters*, September 17, 2012.
14 Gregory Meyer and Javier Blas, "Traders Cause Cotton Chaos with Bulk Deliveries," *Financial Times*, September 25, 2011.
15 United States CFTC, *Rule Enforcement Review of ICE Futures US* (Washington, DC: CFTC Division of Market Oversight, July 2014).

16 See the Options Guide, "Selling (Going Short) Cotton Futures to Profit from a Fall in Cotton Prices." http://www.theoptionsguide.com/cotton-futures-selling.aspx. A profit of US$2,300 could be made in a split second through closing out one short position on a 50,000-pound cotton contract for future delivery. This tidy and instant profit would surpass the total annual returns on seed cotton to at least half a dozen households near Geita, Tanzania, for example. And no cotton necessarily changes hands in real everyday e-trading scenarios involving many hundreds or thousands more cotton contracts.
17 VE News, "ICA Tells Vietnam to Play by Rules," Vietnam News Summary, August 28, 2012.
18 See WTO, *Cotton: Background Paper by the Secretariat* (Geneva: WTO Secretariat, 2015).
19 Thomas J. Bassett, "Capturing the Margins: World Market Prices and Cotton Farmer Incomes in West Africa," *World Development* 59 (2014): 408–21.
20 Ian Scoones, "Cotton in Crisis: The Limits of Liberalization," *Zimbabweland*, November 16, 2015.
21 See in particular Adam Sneyd, "When Governance Gets Going: Certifying 'Better Cotton' and 'Better Sugarcane,'" *Development and Change* 45(2) (2014): 231–56.
22 BioRe Tanzania, *Internal Control System Manual* (Mwamishali: bioRe Tanzania, 2006).
23 See the Maam Samba cooperative, https://fr-fr.facebook.com/MaamSamba/. And BTC, *Fair and Sustainable Trade in Senegal* (Brussels: Trade for Development Centre, 2011).
24 David Vogel, *Trading Up: Consumer and Environmental Regulation in a Global Economy* (Harvard: Harvard University Press, 2009): 259.
25 Tim Smedley, "Better at Scale," *The Cotton Conundrum: Securing the Future of a Very Important Fibre* (London: Green Futures, Better Cotton Initiative, Cotton Incorporated, IDH The Sustainable Trade Initiative, John Lewis and Solidaridad, 2013): 10–11.
26 Reinier de Man, *Promoting Sustainable Cotton Production in West Africa: Potential Supply Chain Strategies: Report to UNEP and FAO* (Leiden: Reinier de Man Sustainable Business Development, 2006). See also the update on the system produced by the BCI itself. BCI, "Better Cotton Standard

System" (Geneva: BCI, 2015). http://bettercotton.org/about-better-cotton/better-cotton-standard-system/. The BCI principles encourage farmers to: minimize the impact of crop protection practices; use and care for water efficiently; care for soil health; conserve natural habitats; preserve the quality of the fiber; and promote decent work. To advance these principles, the BCI supports capacity building but does not train farmers directly. Its assurance program measures performance improvements via farmer self-assessments, credibility checks, and the independent verification of general compliance. Unlike organic and fair-trade systems, it does not, via independent and accredited third parties, certify the compliance of every individual farm in the system with its principles. Rules for the production of cotton that can be certified organic or fair trade are demonstrably broader and deeper than the BCI principles. But the BCI does encourage the enhancement of chain-of-custody documentation in the interest of improving the traceability of "better" cotton, even if its supporters remain unwilling to pay more for the fiber. Those that source the most "better" cotton earn the right to communicate their ostensibly "better" commitments. Ikea prominently did so in a video posted on YouTube in 2013: https://www.youtube.com/watch?v=AubjTw5i26c. Here, the biggest players earn the right to pay pipers to call the sustainability tune and convince consumers that they are committed to doing better. And the BCI only monitors sustainability "progress"in relation to the conventional market, not relative to other certification systems – such as organic – that have been demonstrably linked to more substantive sustainability innovations. See the Textile Exchange, *Organic Cotton Market Report 2014* for more on the latter.

27 Simon Ferrigno, "Cotton Scape," *The Cotton Conundrum: Securing the Future of a Very Important Fibre* (London: Green Futures, 2013): 2–3.

28 Cotton Incorporated, *Life Cycle Assessment of Cotton Fibre and Fabric* (Cary, NC: Cotton Incorporated, 2012).

29 Water use has been a particularly controversial topic. Ground and surface water withdrawals for cotton have had significant impacts on the availability of potable water in numerous drought-prone cotton zones. Strikingly, one rigorous study found that 84 percent of the water footprint of cotton

consumption in the European Union is located outside of Europe, particularly in India and Uzbekistan. See that same study for a rigorous and still relevant approach to assessing the water footprint of cotton. A. K. Chaplain et al., "The Water Footprint of Cotton Consumption: An Assessment of the Impact of Worldwide Consumption of Cotton Products on Water Resources in the Cotton-Producing Countries," *Ecological Economics* 60 (2006): 186–203.

30 John Neal, "The Task of Protecting India's Child Cotton Pickers," *BBC*, 23 February 2014.

31 See the blog posts of the Cotton Campaign: http://www.cottoncampaign.org/.

32 On the latter, see Helen Goworek, "Social and Environmental Sustainability in the Clothing Industry: A Case Study of a Fair Trade Retailer," *Social Responsibility Journal* 7(1) (2011): 74–86. Lisa Richey and Stefano Ponte's broader work on the limits and prospects for philanthropic approaches to ethical consumption and cause-related marketing is also instructive. See their book *Brand Aid: Shopping Well to Save the World* (Minneapolis, MN: University of Minnesota Press, 2011).

33 John Kenneth Galbraith, *American Capitalism: The Concept of Countervailing Power* (Boston, MA: Houghton Mifflin, 1952).

34 The rise of alternative approaches to cultivating cotton has challenged the ways that the industry has traditionally understood the "quality" of cotton. The latter topic has been subject to its own geopolitics over the years. The physical grading of cotton once involved the manual pulling of the fiber to evaluate its staple length, strength, color, and "trash" matter content. Over the past decades, a machine that the US government developed in collaboration with US industry – the so-called 'High-Volume Instrument' – upended manual grading and generated considerable international politics related to the science of grading. For more on this, see Amy A. Quark, "Scientized Politics and Global Governance in the Cotton Trade: Evaluating Divergent Theories of Scientization," *Review of International Political Economy* 19(5) (2012): 895–917. While state control over the classification of cotton continues to generate international politics that the ICAC aims to smooth, global political action has heated up in another area of science: the science of cotton production methods. Control over this

science is now the stuff of serious global politics. Approaches rooted in conventional agronomy, and those wedded to alternative agroecology and organic science, are subject to increasing political contest. As such, the transnational politics of the science of cotton has become more diversified than simple yet meaningful confrontations over downstream classification. The debates that envelop contending scientific approaches to production methods on the land are where the global politics of cotton quality are most dynamic.

35 Ruth Sullivan, "Rana Plaza Companies 'Not Doing Enough,'" *Financial Times*, May 12, 2014.

36 Michael Peel and Barney Jopson, "Cambodia Unrest Stokes Retail Fear," *Financial Times*, January 8, 2014.

37 See William Ridley and Stephen Devadoss, "Analysis of the Brazil–USA Cotton Dispute," *Journal of International Trade Law and Policy* 11(2) (2012): 148–62. And also John Baffes, "Cotton Subsidies, the WTO, and the 'Cotton Problem,'" *The World Economy* (2011): 1534–56. My article in *Third World Quarterly* cited above puts efforts to end US subsidies in a broader political economy and development context.

6 BEYOND THE DIRTY WHITE STUFF

1 Agroecology.org, *Principles of Agroecology and Sustainability*. http://www.agroecology.org/Principles_List.html.

2 Sustainable Cotton Project, *Footprint Calculator* (Winter, CA: Sustainable Cotton Project). http://www.sustainablecotton.org/footprint_calculator/growers/.

3 WWF-India, *Cutting Cotton Carbon Emissions: Findings from Warangal, India* (Surrey: WWF-UK and Marks & Spencer, 2013).

4 Nia Cherrett et al., *Ecological Footprint and Water Analysis of Cotton, Hemp and Polyester* (Stockholm: Stockholm Environment Institute, 2005).

5 See ICAC (2015) and Textile Exchange (2015). The latter report is the source for the statistic in the previous paragraph on the growth of GOTS-certified facilities.

6 Cotton Analytics, "Response to Neil Young, Protect Earth Campaign." http://cottonanalytics.com/97/.

7 Marilyn Waring, *Counting for Nothing: What Men Value and*

What Women are Worth, 2nd edn (Toronto: University of Toronto Press, 1999).

8 Joseph Stiglitz, Amartya Sen, and Jean-Paul Fitoussi, *Mismeasuring Our Lives: Why GDP Doesn't Add Up: The Report of the Commission on the Measurement of Economic Performance and Social Progress* (New York: The New Press, 2010).

9 Subramanian Senthilkannan Muthu, *Assessing the Environmental Impact of Textiles and the Clothing Supply Chain* (Cambridge: Woodhead Publishing, 2014).

10 See Paul Collier on fair trade and poverty in *The Bottom Billion: Why the Poorest Countries are Failing and What Can Be Done About It* (Oxford: Oxford University Press, 2007).

AFTERWORD: A LEARNER IN THE WORLD COTTON ORDER

1 Naomi Klein, *No Logo: Taking Aim at the Brand Bullies* (Toronto: Knopf, 2009).

2 William Greider, *One World, Ready or Not: The Manic Logic of Global Capitalism* (New York: Simon & Schuster, 1998).

3 Paolo Freire, *Pedagogy of the Oppressed*, 30th edn (New York: Bloomsbury Academic, 2000).

4 Gumisai Mutume, "Hope Seen in the Ashes of Cancun," *Africa Recovery* 17(3), October 2003. http://www.un.org/en/africarenewal/vol17no3/173wto.htm.

5 Geoffrey York, "B. C. Professor Ruffles Feathers by Spotlighting Africa's Data Problems," *The Globe and Mail*, November 13, 2013.

6 Adam Sneyd, *Cottonundrum*. http://cottonundrum.blogspot.ca/.

7 For a summary of my thoughts on the limits of framing Africa's cotton 'problem' primarily as a trade issue, see Adam Sneyd, "Beyond the Poverty of Poverty Reduction," *African Arguments*, December 14, 2015. http://africanarguments.org/2015/12/14/beyond-the-poverty-of-poverty-reduction-the-case-of-cotton/

8 Simon Cross, "K2, Cotton and Mountaineering." https://simoncross.wordpress.com/2010/03/17/nick-mason-organic-cotton-k2-and-mountaineering/

9 Adam Sneyd, *Governing Cotton: Globalization and Poverty in Africa* (Basingstoke: Palgrave Macmillan, 2011).

10 WTO, "Plenary Session 15: Reducing Trade Costs in the Cotton Value Chain," *5th Global Review of Aid for Trade*, July 2, 2015. https://www.wto.org/english/tratop_e/devel_e/a4t_e/global_review15prog_e/sessions15_e.htm.

Selected Readings

If you want to learn more about cotton, there are plenty of fantastic sources that can help you to do so. But before you set out on your quest, it is imperative that you follow one overarching rule. Simply put, make sure that you throw your net as widely as possible. Do not cut yourself off from potential insights simply because they do not originate within academia or dovetail with your own politics. Universities do not hold a sole claim to producing rigorous knowledge about this global resource. So if you are willing to embrace diversity and controversy, it is important to look beyond the ivory tower.

And when you do look within it, the same lessons on pluralism apply. You should consult and engage with university-based scholarship on cotton that has its origins in a wide range of academic disciplines. But do not stop there. If this book has not totally killed off any desire that you might have had at the outset to know more about what goes into your clothes, several recently published books on textiles, garments, and fashion should be your guide. Alternatively, if you are fed up with cotton, there are a few books on broader political economy themes that might keep you engaged with the global challenges that make up our lives. The latter might give you some insight into the field within which my contributions are situated and also offer you several clues regarding the political economists and development thinkers that have most influenced my approach to this topic.

Inter-state, international organizations offer a wealth of potential intellectual riches. Reports from the International Cotton Advisory Committee are must-reads. The ICAC Secretariat consistently produces information on developments in cotton production, marketing, trade, and policies. The writings of José Sette, the current executive director, and of his predecessor, Terry Townsend, should be carefully studied. The latter, now a self-described cotton industry consultant, writes opinions on cotton on his website that address matters of pressing business concern. Additionally, the members of the ICAC Expert Panel on the Social, Environmental and Economic Performance of Cotton have individually and collectively authored informative studies and reports.

Beyond ICAC, several other international and nongovernmental organizations have produced significant outputs on cotton over the past fifteen years. Over that time, the World Trade Organization has maintained a particularly strong internet presence. On the WTO website, it is easy to find information on cotton-linked trade disputes and also on the trade negotiations that have pertained to the trade and development aspects of cotton. The *BRIDGES Weekly Trade News Digest*, published by the International Centre for Trade and Sustainable Development, has accessibly covered, synthesized, and analyzed many of the key developments at the WTO. Also of note are the publications associated with the European Union cotton "partnership" with the inter-state grouping known as the African, Caribbean, and Pacific countries. The EU-ACP cotton partnership, launched in the broader context of attention to the cotton "problem" and Africa at the WTO, facilitated the production of several serious reports on cotton, trade, and development.

The latter topics also continue to animate many of the

influential outputs of international organizations that seek to improve agriculture and foster industrialization. As such, they bear upon cotton and should be read. The UN Food and Agriculture Organization's flagship *The State of Food and Agriculture 2015* is as good a place as any to start. So too are the marquee *Trade and Development* reports issued annually by the United Nations Conference on Trade and Development. And if you are interested in food security or the fiber/food trade-off, the collaborations of the UNCTAD with the United Nations Environment Programme are an excellent first point of reference. The pathbreaking *Organic Agriculture and Food Security in Africa* (New York and Geneva: UNEP and UNCTAD, 2008) report offers a comprehensive overview of the ways that alternative approaches to agriculture can enhance food security. The parallel challenge of industrialization has also received extensive coverage. The United Nations Economic Commission for Africa's annual *Economic Reports on Africa* have of late been particularly noteworthy in this area.

Many individuals that serve international, national, and nongovernmental organizations continue to publish serious articles, working papers, and public information on cotton. Of note here are the voluminous research outputs of World Bank economist John Baffes. His Bank colleague Paul Brenton has also authored numerous papers and reports that can help readers to enhance their contextual understandings of the challenges facing developing countries that produce cotton. Claire Delpeuch, now at the Organization for Economic Cooperation and Development, has authored and co-authored several peer-reviewed articles on the economics of cotton in the African context that warrant attention. Turning to the United States Department of Agriculture, the cotton specialists at the Economic Research Service include Leslie Meyer and Stephen

MacDonald. Meyer and MacDonald co-author the USDA's *Cotton and Wool Outlook*. Their reports are widely regarded as key sources of information on the latest developments in the global industry. As regards luminaries in civil society, Gawain Kripke, Sally Baden, and Kevin Watkins have authored influential reports and blogs on the challenges of cotton in Africa. And in the lead-up to the WTO's 2005 Hong Kong Ministerial, Eric Hazard organized and edited a book on Africa's cotton problems.

Beyond these big names, the top tier of non-ivory tower sources also includes the periodic reports and public information produced by several non-state organizations. Textile Exchange, the organic cotton and sustainable textile promoter, publishes exceptional information on the organic cotton market in its *Organic Cotton Market Report*. The International Cotton Association, the Fairtrade Labelling Organization, and the Better Cotton Initiative also each produce reports and public information that offer keen insights into the business of cotton today.

Turning to academia, two scholars have recently published histories that have advanced scholarly understandings of the place of cotton in our world. The University of Warwick's Giorgio Riello and Harvard's Sven Beckert have taken historically oriented scholarship on cotton to the next level. Riello's *Cotton: The Fabric that Made the Modern World* (Cambridge: Cambridge University Press, 2013) reorients the discussion of cotton and industrialization. He draws needed attention to the contributions that Asian innovations and technologies made to industrialization first in England and then in the rest of Europe. As such, Riello does for cotton specifically what John Hobson did for Asia more generally in his groundbreaking book on the *Eastern Origins of Western Civilization* (Cambridge: Cambridge University Press 2004). Sven Beckert's book

Empire of Cotton: A Global History (New York: Alfred A. Knopf, 2014) is now universally regarded as a work of high scholarship on the linkages between cotton and the world economy in the nineteenth century. Beckert consistently and convincingly links the production of cotton to the production of violence, and his book was nominated for the Pulitzer Prize in History.

Other essential readings on the history of cotton tend to focus not only on cotton in a world context, but also on the roles that the industry played in the development of particular regions or countries. For instance, Douglas Farnie and David Jeremy's sweeping edited collection *The Fibre that Changed the World: The Cotton Industry in International Perspective, 1600s–1990s*, (Oxford: Oxford University Press, 2004). In that volume, Farnie and Jeremy first consider the global situation of cotton and then devote the subsequent two-thirds of their book to developments in the United States and Europe, and then to the history of the industry in Asia. Similarly, Alan F. Isaacman and Richard Roberts embrace a regional approach in their co-edited collection *Cotton, Colonialism, and Social History in Sub-Saharan Africa* (Portsmouth: Heinemann, 1995). Isaacman and Roberts offer a compelling history of the brutal life-altering and life-ending realities that were associated with the introduction of cotton cultivation across Africa south of the Sahara. Their fascinating colonial case studies are hard-hitting and even harder to top. But two further works on the histories of particular countries nonetheless warrant special attention. Thomas J. Bassett's engaging study of *The Peasant Cotton Revolution in West Africa: Cote d'Ivoire, 1880–1995* (Cambridge: Cambridge University Press, 2001), and Richard W. Bulliet's intriguing *Cotton, Climate and Camels in Early Islamic Iran: A Moment in World History* (New York: Columbia University Press, 2009).

Academic economists and geographers have also shed considerable light on aspects of the global industry. Colin Poulton's research on cotton production and marketing systems in countries where smallholders are the primary source of cotton warrants sustained engagement, as does the work of his numerous collaborators on cotton and textiles: a grouping that prominently includes the legendary Brian Cooksey. Turning to the geographers, William Moseley and Leslie Gray offered a solid spatial analysis of Africa's cotton problems in their co-edited volume *Hanging by a Thread: Cotton, Globalization, and Poverty in Africa* (Ohio: Ohio University Press, 2008). Other notable geography research has focused on the spaces where cotton is produced in relation to the global cotton chain. Marianne Larsen's doctoral dissertation, completed at the University of Copenhagen, stands as an excellent example of the contributions geographers have made to understanding cotton.

From sociology, and also from political science, Peter Gibbon's numerous research articles offer incredible insight into the lives of smallholder cotton farmers. His co-authored volume with Stefano Ponte, *Trading Down Africa, Value Chains and the Global Economy* (Philadelphia: Temple University Press, 2005), continues to stand as the essential reference for understanding commodities in the African context. The sociologist Amy Quark has taken a very different point of departure. In her book *Global Rivalries: Standards Wars and the Transnational Cotton Trade* (Chicago: University of Chicago Press, 2013), she exposed the politics of science and rule making that cloak the grading of cotton quality and infuse the related and highly technical dispute settlement procedures. Koray Çalişkan, a political scientist, offered a broader sociological orientation on the creation and re-production of the global cotton market as a whole in his book *Market Threads: How Cotton*

Farmers and Traders Create a Global Commodity (Princeton: Princeton University Press, 2010). To do so, Çalişkan drew extensively upon field research conducted in Egypt, Turkey, and elsewhere.

To get a better sense of the fiber in relation to the downstream industry, Pietra Rivoli's *The Travels of a T-Shirt in the Global Economy: An Economist Examines the Markets, Power, and Politics of World Trade* (Hoboken, NJ: Wiley Publishers, 2005) is essential reading. This book has been updated multiple times since its original publication and remains the go-to introductory source on this topic. A fantastic complement to Rivoli's book has recently emerged in the form of *Clothing Poverty: The Hidden World of Fast Fashion and Second-Hand Clothes* by Andrew Brooks (New York: Zed Books, 2015). Brooks expertly takes his readers through the darker side of the downstream industry. *Clothing Poverty* was written not only for political elites who have an interest in making industrialization work for development: Brooks speaks directly to consumers of fast fashion and poses troubling questions. And the dumping of used clothing in developing countries continues to compound the problems that Lynn K. Mytelka identified in her classic article on the "The Unfulfilled Promise of African Industrialization," *African Studies Review* 32(3) (1989): 77–137.

Beyond cotton, several political economy readings offer keen insight into how the world works and does not work. So if you are up to the challenge of thinking more deeply about how the world in which you live is ordered by empowered people, Robert W. Cox's *Approaches to World Order* (Cambridge: Cambridge University Press, 1996) is a fantastic starting point. To learn more about the role of commodities in social change, the writings of self-described "dirt" researcher Harold Innis are foundational. The collection compiled by Daniel Drache in *Staples, Markets and*

Cultural Change: Selected Essays of Harold A. Innis (Montreal and Kingston: McGill-Queen's University Press, 1995) is especially relevant for readers in developing countries that continue to depend on raw material exports. And if you are interested in knowing more about how the world order is changing and remaining the same, check out any of the volumes in the *Globalization and Autonomy* series (Vancouver: UBC Press) edited by William D. Coleman and colleagues. Also, for an enriching up-to-date overview of the intersection of agricultural livelihoods, rural development, and political economy, see Ian Scoones, *Sustainable Livelihoods and Rural Development* (Warwickshire: Practical Action Publishing, 2015).

Finally, to understand the intersection of politics and economics in our everyday lives, there is no better source than John Kenneth Galbraith. *The Essential Galbraith* (Boston, MA: Houghton Mifflin, 2001) is a notable collection of the most enduring writings of the twentieth century's towering economist. Cambridge economist Ha-Joon Chang offers incredibly accessible reading on the interface of politics and economics in the twenty-first century in his book *Bad Samaritans: The Myth of Free Trade and the Secret History of Capitalism* (New York: Bloomsbury, 2007). Chang expands his focus on the politics of trade to the politics of capitalism more generally in his now classic *23 Things They Don't Tell You about Capitalism* (New York: Bloomsbury, 2010). And if those fantastic books do not hold your interest in political economy, please consult John Perkins's updated bestseller *The New Confessions of an Economic Hitman* (San Francisco, CA : Berrett-Koehler Publishers, 2016). Unfortunately the kinds of nefarious dealings Perkins details did not end with the Cold War. See *The Looting Machine: Warlords, Oligarchs, Smugglers, and the Theft of Africa's Wealth* (New York: Public Affairs, 2015) by Tom Burgis. He offers the real deal on the current politics of underdevelopment in economies that are overly dependent on export commodities. For another informative take on developments in this area, see Ian Taylor, *Africa Rising? BRICS – Diversifying Dependency* (Suffolk: James Currey, 2014).

Index